'A wonderfully lyrical, yet scientifically rigorous and wide-ranging exploration of the way trees enrich our lives. Importantly, Jake Robinson identifies the biological and psychological imperatives to properly restore our lost forests, and the myriad life forms they encompass. Go sit under a tree and read this book today!'

—Mike Bossley, conservationist and former director of Greenpeace Australia

'*Treewilding* is a major addition to literature on inter-relationships between people, trees, and forests. It takes a broad sweep through environmental and cultural histories of the forest and impacts of deforestation. The author then looks to future restoration with a context of social and ecological values. The book is an important, innovative, and hugely stimulating contribution that demands to be widely read.'

—Professor Ian D. Rotherham, Advanced Wellbeing Research Centre, Sheffield Hallam University

'I very much enjoyed reading this belter of a book. Timely and important, it is both a paean to the power and beauty of trees and a call to arms. Let trees be thy teachers and this book be your guide.'

—Dr George McGavin, entomologist, broadcaster and President of the Dorset Wildlife Trust

'Electrifying and insightful! With personal anecdotes and scientific rigour, Robinson charts our course to a more symbiotic existence with forests. Like the dendritic branching structures it so beautifully describes, *Treewilding* takes us on a journey through time, revealing the relationship between humans and trees is both ancient and complex - one that we break at our peril.'

—Tim Jarvis, adventurer and environmental scientist

'A characterful and informative exploration.'

—Tristan Gooley, author of *How to Read a Tree*

'There has never been a more important time for the world to understand trees.'

—Bruce Pascoe, author of *Dark Emu*

'*Treewilding* is captivating. Each one of us senses a powerful connection with the nature that shines all around us, even if we cannot articulate it. Jake does so beautifully in this wonderful book.'

—Ben Goldsmith, environmentalist and author of *God is an Octopus*

'Robinson depicts precisely and gracefully how trees and forests are interconnected with each other and with the world, from the smallest to the largest scale, in imaginable and unimaginable ways. The book is, while being scientifically thorough, an invitation to sense, feel and appreciate this interconnectedness ... it is impossible to be unenthusiastic about this wonderful book.'

—Hanna Bjørgaas, author of *Secret Life of the City*

TREEWILDING

JAKE M. ROBINSON is a British microbial and restoration ecologist currently residing in Australia. He earned a PhD from the University of Sheffield in 2021. His research interests span microbes, ecosystems, social equity issues and strategies to conserve and restore nature. *Treewilding* is his second book.

www.jakemrobinson.com

Find Jake on X (Twitter) @_jake_robinson and @MicrobeBook

Jake's first book was *Invisible Friends*. It's all about how microbes shape our lives and the world around us – often in beneficial ways!

TREEWILDING

Our Past, Present and Future Relationship with Forests

JAKE M. ROBINSON

PELAGIC PUBLISHING

First published in 2024 by
Pelagic Publishing
20–22 Wenlock Road
London N1 7GU

www.pelagicpublishing.com

Treewilding: Our Past, Present and Future Relationship with Forests

All illustrations and photographs by the author unless otherwise stated.

https://doi.org/10.53061/GBOY6819

A CIP record for this book is available from the British Library

ISBN 978-1-78427-480-1 Hbk
ISBN 978-1-78427-481-8 ePub
ISBN 978-1-78427-482-5 PDF
ISBN 978-1-78427-512-9 Audio

Cover design by Jennifer Smith

Typeset in Minion Pro by S4Carlisle Publishing Services, Chennai, India

Contents

Acknowledgements

Writing this book has been a journey filled with inspiration, challenges and growth. I extend my heartfelt gratitude to the individuals and groups who made this endeavour possible. I am deeply thankful to my wife, Kate, whose unwavering support and encouragement help fuel my passion for translating science into accessible narratives. Kate also tolerates my impromptu tree-climbing escapades in the garden and parks and coming home covered in dirt and smelling of damp wood after a long day surveying bats in the treetops for work. On this note, I also dedicate the book to my cat, Foxtrot, whose love of climbing trees is equal to my own (even though we watch the perching bluetits for very different reasons – and don't worry, we keep her on a leash).

To my family, friends and mentors who provided support during the writing process, thank you. I also extend my gratitude to the dedicated conservationists, researchers, scientists, historians, artists and philosophers whose work and passion form the foundation of this book. Their commitment to advancing knowledge in their respective fields and caring for the planet has been a constant source of inspiration.

A special thanks to those who shared their personal experiences and insights, adding a unique and authentic dimension to the book. Their stories have illuminated the human connection to the topics explored. Many thanks to Dr Harry Watkins, Prof. Forrest Fleischman, Prof. Qing Li, Keith Bradby, Dr Richard Broughton, Klaus Lotz (and family), Dr Alex Schnell, Dr Charnett Chau, Dr Helen Whitehead, Prof. Line Nybakken, R.J. Ranjit Daniels, Joe Middleton and James from the Woodland Trust, Prof. Toby Kiers, Shelley Castle, Prof. Bruce Pascoe and Brian McHugh for their expertise and contributions, enhancing the depth and accuracy of the scientific content presented. Their collaboration has truly enriched the narrative. Thanks to Dr Robin Taylor for the stimulating conversations we have on our podcast *Interconnected* – it all contributes to the ecology of mind that shapes the narrative.

Thanks also to the many collaborators on the studies I reference throughout the book, including Prof. Paul Brindley, Prof. Suzanne Mavoa, Dr Carlos Abrahams, Assoc. Prof. Martin Breed, Prof. Anna Jorgensen, Dr Craig Liddicoat, Christian Cando-Dumancela, Prof. Miles Richardson, Dr Ross Cameron, Dr Brenda Parker, Dr Kim O'Donnell, Prof. Nicole Redvers, Dr Jacob Mills, Dr Nick Gellie, Araceli Camargo, and Prof. Sue Ishaq, Dr Peter Harrison, Dr Rachel Antwis, Dr Ravin Poudel, Dr Danielle McCarthy and Prof. Chris Daniels. My appreciation also goes out to those who have inspired the book, but to whom I may not have spoken about it: Prof. Suzanne Simard, Prof. Robin Wall Kimmerer, Prof. Robert Sapolsky, Prof. Richard Dawkins, Prof. John Cryan, Prof. Susan Prescott, Dr Monica Gagliano and the late Profs Aldo Leopold and Lynn Margulis.

My sincere appreciation also goes to my publisher, Pelagic Publishing, for their invaluable insights and guidance throughout the writing process. Thanks to Nigel Massen for his willingness to take this project on and Sarah Stott for her marketing advice. I also express my appreciation to David Hawkins for his meticulous editing, ensuring clarity and coherence in the manuscript. His keen eye and thoughtful suggestions have significantly improved the overall quality of the writing.

Finally, to the readers who embark on this journey, I extend my deepest appreciation. Your curiosity and engagement make the endeavour of sharing knowledge truly meaningful.

Thank you to all who played a part in bringing this book to life.

INTRODUCTION

Seeing the Woods for the Wood

"If we are looking for models of self-sustaining communities, we need look no further than an old-growth forest. Or the old-growth cultures they raised in symbiosis with them."
—Robin Wall Kimmerer

Seeing the woods (complex ecosystems) for the wood (timber)

I'm sitting on a log, watching the crispy leaves from a nearby forest dancing in the breeze; winter's chilly promise is just around the corner. The rocks are hugged by a fluffy cloak of moss, their edges decorated by algae, like glistening jade necklaces. The log's bark is covered in sunburst lichens and a stand of curly bracken embellishes the margins in the distance. There are no mature trees, but tree saplings are dotted across the landscape. They were planted here by humans, and now they compete and maybe even cooperate as their young branches and leaves with tiny chlorophyll parcels reach for the sky. Some saplings will hopefully survive and grow into a flourishing community. Many dangers lie ahead for them, but in the right conditions nature is remarkably resilient. This is the start of a forest restoration project – a local story of hope. And what better symbol of hope than the seed of a tree, fragile yet brimful of potential?

At a global scale, we teeter on the precipice of unrivalled biodiversity loss and climate chaos, so restoring forests (and other) ecosystems is vital. But the truth is that forest restoration often fails. It fails if we don't consider the livelihoods of the people who live on and care for the land. It fails if we neglect to apply ecological knowledge or promote Indigenous leadership. And it fails because we often view trees as commodities or wooden street ornaments instead of complex systems. Underscoring all these issues is our growing disconnection from each other and the land beneath our feet, particularly in 'high-income' societies. It's safe to assert that we cannot effectively conserve and restore nature if our worldview is predicated on control and convenience rather than reciprocity and belonging.

Two spiritual dangers

As Aldo Leopold famously said, "There are two spiritual dangers in not owning a farm. One is the danger of supposing that breakfast comes from the grocery, and the other is that heat comes from the furnace."[1] The same concept applies to paper and timber. How often do we think about where paper comes from? As our connection with nature weakens, is there a spiritual danger in supposing that paper comes from the stationery shop and not the forest – a complex ecosystem? Having written a book on restoring forests, I feel the need to acknowledge the sacrifice of nature in facilitating the delivery of these words. Each page of this book is a gift from nature. Each page was once part of a functioning ecosystem, a forest, a dynamic community of life.

When we disconnect the product from its source, we risk devaluing the interdependence of life. Indeed, many of us now live in a world of luxury where convenience is experienced at the click of a button. But it's important to take the time to appreciate our evolutionary roots, the roots of our convenience and the roots of our destruction. It's also important to understand the multifaceted values of trees and forests, support the rich bounty of knowledge of different cultures and build on restoration successes. By doing so, we can shape the future in wonderfully positive ways and cultivate a deeper respect and reverence for the natural world.

What exactly is a tree?

And why is it important to restore trees in our landscapes? A tree is myriad things, transcending the boundaries of objectivity. A tree is a subject and a living community. A paper mill owner may see a tree as a resource, pulp-to-be. Living trees can be ancient. They've lived through medieval history, quietly witnessing our tragic and halcyon days. Some trees burst from their seeds and anchored their roots several millennia ago. Indeed, some living trees are nine times older than the first paper mill established in Nuremberg, Germany, in 1390.[2] These trees continue to inhale our exhalations today as we inhale theirs. Carl Sagan famously said that it's a "marvellous cooperative arrangement ... a planet-wide mutual mouth-to-stoma resuscitation"![3]

Where does the word 'tree' come from? The answer lies in thousands of years of history spanning the entire European continent. We know the word 'tree' was uttered in England 500 years ago by the great playwright William Shakespeare. Yet in tenth-century England the word typically appeared as *treow*, a derivative of the Germanic *treuwaz*. This stemmed from another word, *dreuom*, around 3,000 years ago. But we can travel back further still, to 4,000 years ago and to the land around the Caspian Sea, where the etymological trail ends. Peoples of the Kurgan culture spoke Proto Indo-European – the oldest predecessor of Modern English – and used *dóru* to describe a tree. This word carried meanings such as 'steadfast', 'hard' and 'strong'. We somehow got from *doru* to 'tree', but it's easy to understand the reasons behind the meaning – trees certainly conjure up images of strength, hardiness and resilience in my mind, along with beauty, serenity and wisdom.

Just one tree species can provide vital habitat for hundreds of animals: insects, birds, bats and reptiles. Trees are part of an ecological community. But trees are also a community in themselves – 'holobionts' – a host plus many microbial partners that work together as a functioning ecological unit. Swathes of bacteria, fungi, archaea, viruses, algae and protozoa rely on the trees, and the trees depend on them. It's an intimate relationship spanning multiple kingdoms of life, from the visible to the almost imperceptible.

To many Indigenous Peoples worldwide, a tree is a community member. In fact, in many cultures, all elements of nature are animate subjects, not objects.[4] Indeed, the trees are considered sentient. They inspire art and provide gifts in the form of medicines, food and tools. Indigenous Peoples' deep connection with trees and the respect shown to them is arguably a profound omission from capitalist thinking. This has consequences for biodiversity, the climate, our lives, our kin and future generations. Having a deep sense of respect for trees is indicative of a broader cultural and economic system that prioritises long-term sustainability and collective wellbeing over short-term gains and individual profit – and the opposite is also true.

A tree, like other plants, is a sensitive organism. Indeed, it has senses. A tree can perceive, assess, learn, remember, and it can solve problems. A tree is able to make decisions and potentially communicate with other trees by actively acquiring information from its environment. A tree does not possess a brain or neurons, but it does have competencies that resemble cognition in a perceptual sense.[5] Although contested, some evidence suggests a tree can protect and nurture another tree when it signals for help.[6] In one sense, a tree has a family and a family tree.

Trees are instrumentally valuable to humans. In urban environments, they provide shading and cooling. They regulate nutrients and stormwater. They remove pollution and emit health-promoting microbes into the air. They provide food in nuts and berries. A tree is also aesthetically pleasing; its fractal patterns soothe the human mind by stimulating the so-called rest and digest part of our central nervous system – the parasympathetic nervous system. Here, a flurry of neurotransmitters called acetylcholine are liberated during vagus nerve stimulation, which calm the heart rate in a profound display of biochemical intricacy.[7] A tree teaches us to play, providing a frame for children to climb and play hide-and-seek and with sticks to build dens. A tree provides a meeting point for convivial picnickers yearning to be closer to nature. A tree provides and endures. Restoring trees restores their vast and unparalleled intrinsic and instrumental value.

What are forests?

And why is it important to restore forests? Forests are myriad things. They're more than conglomerates of wooden structures standing tall. Communities of sub-communities, forests contain multitudes. They're places of belonging. They take many years to establish and grow; to form complexity and interactions; to build resilience, bonds and entangled life forms. Forests are networks. They are labyrinthine wonderlands of life and mystery.

The word 'forest' comes from the Latin *forestis silva* or 'outside woodland'. It was applied to areas of land frequented by royal families for hunting. But they are far more than mere hunting grounds.

Forests are vast arenas of light-catching, life-giving entities, shimmering, entangling, drinking and swaying. They're home to many creatures, great and small (three-quarters of the world's life on land can be found in forests). They're also complex ecological systems in which tree holobionts are the dominant life forms, bound by a rich tapestry of geological and climatic conditions. A subterranean network of roots, mycelia (branches of fungi), microbes and other natural elements may connect the trees in a forest like a vast biological internet. The trees in the forest may even communicate.[9] They might even warn each other of invaders through chemical signalling, and they shuttle nutrients to each other: 'You scratch my bark, I'll scratch yours.'

But trees are not the only plants in a forest ecosystem. Forests can often be divided into three or four layers or 'storeys'. There's the forest floor, which is typically dark, warm and humid. Only a few beams of sunlight may hit the ground, especially in dense rainforests. Large-leaved shrubs, creeping herbs and tree saplings grow in the sporadic patches of sunlight. One level up is the understorey: it's often dark and humid there. This is a tangled network of shrubs, young trees and vines that climb and coil around the veteran trunks, aided by their highly attuned sense of touch. Next is the canopy layer. In temperate forests, this forms the roof over the understorey – a complex maze of leaves and branches. Swathes of creatures call this layer their home or a pitstop since food is abundant. Rainforests have a fourth layer – the emergent layer. This refers to the tops of trees that poke up above the rainforest canopy, which reach close to 91 m into the air. However, some forests can be sparse, light and open with only a scattering of trees: heterogeneous rather than uniformly dense. Yet we must remember that all the woody and leafy layers we can see with the naked eye intimately depend on the tangled complexity that eludes our sight. Indeed, the interactions and diversity above ground are dwarfed by those beneath the forest floor. Therefore, forests are as much invisible as they are visible and restoring them restores a home to myriad organisms.

Forests are viewed, defined and valued through different lenses. The choice of one's lens will depend on one's psychological make-up, cultural upbringing, personal experiences and reflections, genes and the environment itself. These all shape the prism of our worldview. Through these various lenses, forests can be seen as possessing intrinsic value – a value in their own right, independent of human uses. Or indeed, as having instrumental value – a physical and spiritual home for many people, a resource for timber production, an

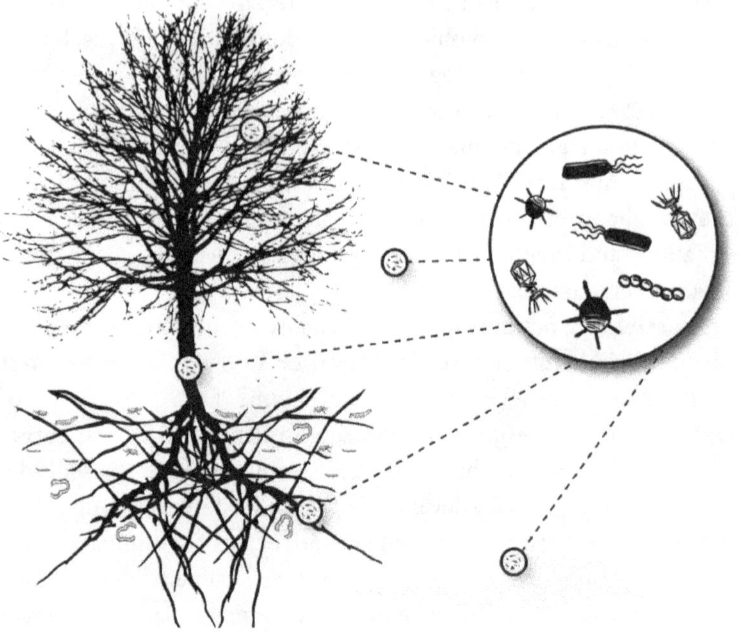

A forest is as much invisible as it is visible.

ecosystem composed of myriad life-giving organisms, a repository for carbon storage, a rich bounty of so-called ecosystems services, a canvas for innovative farming systems, or all the above. Beyond the tangible gifts they give lies a deeper value that's hard to perceive. For in their presence we find peace, a soothing balm to our ever-busier existence. This mélange of definitions reveals that a forest is a treasure chest of complexity, beauty, meaning and life.

What do trees mean to me?

I recall, as a child, spending countless hours in forests. A visceral urge compelled me to be surrounded by the power and awe of trees. Tree climbing was, and still is, a passion of mine. I used to climb trees for fun, sometimes bunking off school with a friend to spend time in the canopy of the local park. We'd sit in the comforting cradle of a large oak, talk about evolution and time travel paradoxes with a healthy sprinkling of incoherent babble and watch the day go by. I now climb trees for pleasure and solace. Sometimes I'm lucky to get paid for this. I find tree canopies to be one of the most soothing places. Perching up there with the squirrels and birds, surrounded by the flittering leaves and diverse lichen and bryophyte communities unseen from the ground – it's calming, educational and inspiring, and the higher you get, the more humbling it becomes. As an ecologist, I regularly ascend tree trunks and scale the branches to check for bats and birds who roost, hibernate and breed in the cracks and crevices. This is all in the name of nature conservation, but the peace, solitude and direct connection with these life-forms are also important to me. If I ever feel disconnected from the rest of nature, which has become all too easy in these digital and urbanised times, climbing a tree, preferably bare footed, swiftly instils in me a renewed sense of reverence and connectedness. To me, the act of restoring a forest is synonymous with restoring meaning. It's about restoring the intricate web of life that makes these ecosystems thrive, but also about re-establishing our connection to the natural world and the meaning it brings to our lives.

Forests under threat

The UK was recently named one of the most nature-depleted countries in the world.[8] In light of this, ecosystem restoration (including forests) should be a national priority. The UK's woodland area is currently 3.2 million ha or 13% cover.[9] However, much of this comprises plantation 'monocultures' that feed the timber industry. The average forest cover of many European countries is around 38%, so the UK is lagging dramatically.[10] Some people think our woodland cover was reduced to 50% by the time of the Iron Age (2,500 BP),[11] after which large-scale human habitation, colonialism and wars further depleted these ecosystems. The tropics alone lost 12.2 million ha of tree cover in 2020 – that's three times the size of the Netherlands.[12] So, despite being myriad things (homes, gift-givers, biodiversity supporters, with both intrinsic and instrumental value), trees and forests are rapidly being diminished to feed our obsession with growth. Yet there are important nuances to this narrative, as you'll discover later in the book.

The integrity of forest ecosystems depends upon an eclectic embroidery of biodiversity, the variety of life in an environment. Microscopic creatures supply trees with nutrients and the ability to converse with the local environment in chemical languages we're only just starting to understand. The forest's animals, plants and microbes contribute to energy flows, disease protection, and soil formation and structure. The tight-knit relationships between these organisms allow the forest to be resilient. Without this complexity, the health of the forests suffers, and their capacity to bounce back from extreme stress, such as climate change, dwindles. Our thirst for convenience at all costs means we often fail to see the *woods* (the ecological and social systems) for the *wood* (the timber).

Restoring forests

Ecosystem restoration is the practice of renewing and restoring degraded or destroyed ecosystems. Think of it as helping nature to heal after a nasty disease (the disease being habitat destruction or pollution). There's an imperative to conserve and restore our forests and other ecosystems globally because, without effective interventions, 95% of land on Earth is projected to be affected by degradation by 2050.[13] This will have unfathomable impacts on biodiversity, climate change and the lives of billions of people. Yet, as mentioned, many ecosystem restoration projects have failed, despite tremendous amounts of money and scientific knowledge being sent their way.

Forest restoration (repairing damaged or destroyed forests) is a growing scientific subdiscipline and an important one in many ways. The discipline is young, and advances in technology (e.g., DNA sequencing and ecoacoustics) are helping us to understand forests on a deeper level. But an all-embracing catalogue of knowledge of how to successfully restore forests and other ecosystems has been held by many ancient cultures for millennia. Despite Indigenous Peoples inhabiting only 22% of the Earth's surface, their communities currently protect 80% of the world's remaining biodiversity. Yet there has been a history of relocating Indigenous communities in the name of nature conservation, and this still happens today. This is not only detrimental to the Indigenous Peoples but also to the biodiversity in the forests and many other precious habitats that they steward.

Tree-planting campaigns have taken off stratospherically in the last few years, with everyone from 'influencers' on YouTube to high-powered CEOs embracing them as revolutionary solutions to the climate and biodiversity crises. However, planting trees where they do not belong, for instance on peat bogs, can harm ecosystems and reduce biodiversity. It can also exacerbate wildfires and climate change. Simply saying "I'm going to put money into planting a trillion trees" often fails to acknowledge holistic options such as fostering community-centred forests and providing support to help Indigenous Peoples defend their lands. This will have a far more significant impact on forest ecosystem restoration.

We need to take an ecological approach to tree planting. We must also take a sociological (human community-centred) approach to tree planting. Otherwise, our children may soon live in a world where 95% of the land is degraded.

A story of reciprocity

This book dives deep into the fascinating, sometimes controversial, but more often hopeful world of forest restoration and science. I truly feel that to understand forest restoration and all its benefits and pitfalls, we must also immerse ourselves in the past and the reasons for the deforestation. As the old adage goes, "Only by understanding the past can we shape our future." Learning from mistakes and building on successes is imperative for us to continue to grow and thrive. To this end, the book begins by unravelling the evolution of forests – how they've risen and how they've fallen. It also considers why they matter, to us and to the myriad life forms we share the planet with. The book then covers the many facets of forest restoration – the good, the bad, the ugly and the hopeful, and ends with unseen, underappreciated and future perspectives. I wrote many parts of the book while walking with a voice recorder in hand through different woodlands of the UK. Through transcribing my experiences, I hope to guide your imagination on a journey through enchanting woodlands while learning about restoration.

There will be musings about evolutionary processes from tree–bird coevolution to epigenetics (how behaviours and environments cause changes that affect how genes work without altering DNA sequences). Evolution can provide insights into adaptations and characteristics that allow nature, including us, to thrive.

Forest restoration is not just ecological. It's also social. And for some, it can be spiritual. Sustaining this nature-healing journey may require a shift in worldview, or at the very least an embracing of other worldviews. It will undoubtedly require societies to foster reciprocal relationships with these life-sustaining communities. *Treewilding* also intends to illuminate a pathway to hope and reciprocity by promoting awareness of underappreciated perspectives. This includes the invisible world, tree senses and memory, Indigenous views and climate change. We can restore our forests and often allow them to restore themselves, just as we can restore our weakening connection with the land. These phenomena are truly interdependent, as humans are very much a part of 'nature'.

The unseen

As a microbial ecologist, I constantly seek ways to convey the importance of the unseen. After all, it's challenging to appreciate what you cannot see. This is why I include a strong emphasis on underappreciated perspectives throughout the book, particularly in the later chapters. We must breach the boundaries of our perceptual world to understand that visible forest biodiversity depends on the unseen realm. Are there lessons of cooperation, interdependence and diversity to learn from our invisible friends? I chat with leading soil fungi expert Toby Kiers to discuss this further, along with the hotly contested 'wood wide web' concept.

Another underappreciated perspective is the notion that trees have cognitive and sensory competencies. In the twentieth century, German biologist Jakob von Uexküll developed a concept called the 'Umwelt'.[14] This describes the sensory realms of different species, that is, their distinct internal model of the world, formed by their perceptions. Different animals have their own unique Umwelt and perceptual tools (sensory organs) that limit what they can sense and, therefore, how they make sense of the world. For instance, humans and other apes have ears. These are a perceptual tool. They allow us to detect sounds from our environment, enabling us to make sense of the world around us. The same goes for our eyes, nose, mouth, skin and so on. But our perceptual tools only allow us to detect a small amount of the immense information out there. A bat, for example, also has ears and can emit and detect sound frequencies that are too high for humans via echolocation. The sound is real; our limited sensory organs just can't detect it. The same goes for other signals in the world, for instance infrared radiation. All warm-bodied organisms emit this, but humans cannot naturally detect it, whereas a pit viper snake has special sensory organs that allow it to easily detect an infrared-emitting rodent (its evening meal) scurrying around in the dark of night. But do trees and other plants have an Umwelt? Do they have perceptual tools that allow them to make sense of the world in unique ways?

The inanimate pages of this book you're reading, now delicate slivers of dehydrated pulp, used to be part of a complex living system capable of responding to and learning from stimuli in its surroundings. Trees (and other plants) can detect signals – chemical, sound and light – and can respond to these signals to 'choose' and make decisions that are most appropriate to ensure their survival. But like the pit viper and the bat, trees do it differently from humans.

It's vital that we brush away the egocentrism that positions humans at the top of a hierarchy of life and cultivate a deeper connection with, and truly respect the complexity of, other life forms. I hope these perspectives strengthen the case for doing so with trees.

Understandably, there is a degree of 'eco-anxiety' in forest restoration as we continue to fight a challenging battle to protect nature. But as Martin Luther King Jr said, "We must accept finite disappointment but never lose infinite hope."[15] There are many examples of successful forest restoration ideas and practices that we should promote to catalyse further positive action. We should also acknowledge the origins and destinies of trees and how they influence our lives. By doing so, we can work towards a life-affirming future for ourselves and our non-human relatives, including those who gifted these pages.

Scattered throughout the book, you'll notice a small tree ring image at the bottom of some pages:

This tree ring indicates that the following page will include a passage about the destiny of trees. These passages are independent of the chapters and include the words 'I Once Was a Tree' in their title.

They're small acknowledgements that many of the products we use in our everyday lives were derived from trees.

Trees and forests are myriad things. They are both visible and invisible. Objects and subjects. Ecological and social. Homes, communities, gift givers and solace providers – and we must band together to protect them. On this note, I hope you enjoy reading *Treewilding: Our Past, Present and Future Relationship with Forests.*

PART I

TREES: THEY RISE, THEY FALL, THEY MATTER

CHAPTER 1

A Brief History of Forests

> *"The forest spoke to my soul in a language I already knew, a distant lullaby from the womb of peace and solitude!"*
>
> —*Angie Weiland-Crosby*

I once watched a video of a biologist teaching schoolchildren about the history of life on Earth. The teacher stretched his arms out wide and said something along the lines of, "Imagine this was a timeline depicting the history of planet Earth, from the tip of my left hand to the tip of my right. Now imagine taking a nail file and lightly stroking the end of my fingernail. The nail dust that falls to the ground represents the duration of human existence." Indeed, the universe is thought to be 13.8 billion years old and Earth 4.5 billion. Yet *Homo sapiens* didn't appear until around 300,000 years ago. However, for this chapter, I'd like you to imagine jumping into a time machine and travelling back long before humans first appeared, to a time when plants first emerged on land. To do this, you'll need to set the time machine dial to 500 million years ago. On the teacher's timeline, this would be a little more than a hand's length.

Your time machine has arrived at 500 million years ago, before dinosaurs roamed the Earth. You step out and see an abundance of green algae in the primordial shallows of coastal waters, but they've also recently evolved to live on dry land.[1] This evolutionary development came with considerable challenges. For one, life on dry land is (no surprises) a less aqueous affair. Desiccation is not favourable for most living entities, particularly those adapted to aquatic habitats. To overcome this fundamental obstacle, a group of early pioneers (the vascular plants) developed anchoring roots, tubular tissues to transport water and a waxy layer to prevent water loss. But this process didn't happen overnight. Indeed, it took tens of millions of years to develop and refine these attributes – a genetic revolution. The number of new genes that emerged when plants first evolved to live on land is far larger than at any other time in plant evolutionary history.[2]

Turn your time machine dial to a few million years ahead, and you'll see that the early terrestrial plants were short and stubby. These plants played a crucial role in paving the way for the emergence of the more complex and towering vegetation we recognise today. However, jump forward to around 390 to 370 million years ago, during the Late Devonian period and take a step outside. Now you'll start to see the first tree-like plants.[3] This transformative period witnessed a remarkable shift in the botanical landscape, as these plants represented a significant departure from the previous vegetation forms. Their advent marked a turning point in the evolution of terrestrial ecosystems. The emergence and diversification of tree-like structures allowed for greater access to sunlight and enabled plants to exploit vertical space more efficiently.

The increased vertical growth also offered opportunities for seed dispersal over greater distances, promoting plant colonisation and diversification across various habitats. The appearance of tree-like plants during this period profoundly impacted terrestrial ecosystems. You'll notice that, unlike the desolate, rocky and somewhat slimy landscape of your first port of call, these new plants have created microclimates beneath their canopies and provide resources allowing diverse creatures, including both invertebrates and vertebrates, to evolve and exploit these new ecological niches.

Branches

It's remarkable to think about how branches manifest across varying scales, spanning the delicate fibres extending from microglial cells in our brain to the intricate network of capillaries that deliver essential nutrients to our cells. Moreover, branches emerge in the extensive underground mycelial networks, intricately connecting plants and facilitating nutrient exchange, as well as in the grandeur of river deltas, where water and sediments converge as they flow into the vastness of the sea. The occurrence of branches across powers of ten exemplifies the interconnectedness and intricate patterns that pervade nature at both the microscopic and macroscopic levels.

The specific genetic mechanisms, cellular processes and environmental interactions involved in the evolution of branching differ across these contexts. However, the formation of branches is typically driven by efficiency and resource distribution. Most branching systems have evolved to optimise the distribution or acquisition of resources, and the same applies to plants and trees. Early rooted plants exhibited branching structures both underground to provide stability and acquire nutrients and above ground to maximise their exposure to sunlight, the life-giving energy source. There's no doubt that branches played a vital role in the evolution of plants and forest ecosystems… and I would go as far as saying all life on Earth.

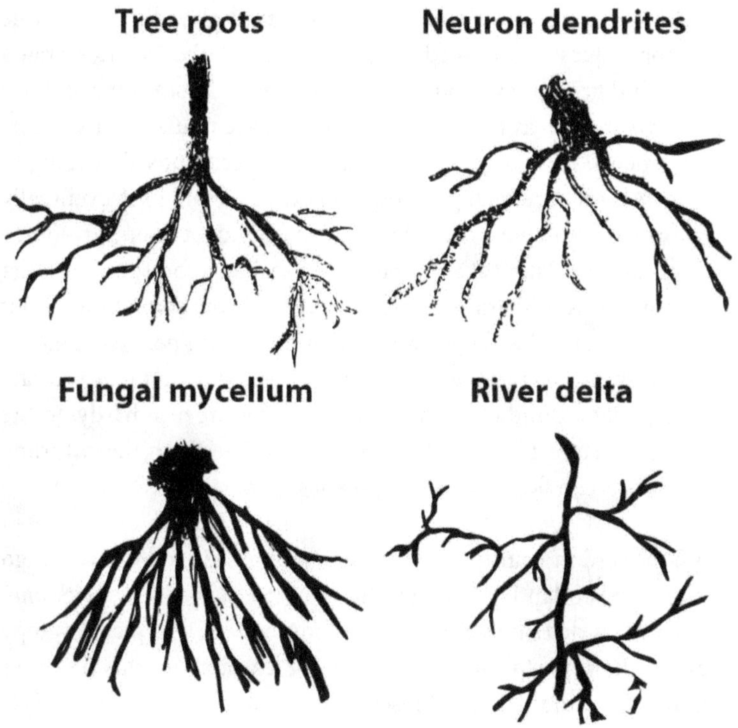

Branches are seen throughout the natural world and were vital in tree evolution.

Back to the time machine

You're now going to 360 million years ago (the Carboniferous period). You'll see that plants have increased in size and have established communities. Whizzing around you are diverse and colossal insects.

Oxygen levels were significantly higher than today, allowing insects to grow to enormous sizes. Dragonflies, for example, had wingspans of up to 75 cm.[4] The Carboniferous period also marked the rise of amniotes, a group of vertebrates that possess amniotic eggs. This evolutionary adaptation enabled them to reproduce on land and freed them from the dependence on water for egg-laying. Early amniotes eventually gave rise to reptiles, which became dominant during subsequent periods. A marvellous array of competition and cooperation drove trees to radiate and diversify. Most species from this period are now extinct, and much of the coal that provides fossil fuels comes from the vast graveyard of these Carboniferous trees. However, a handful of giant Carboniferous tree fern species still live today in the forests of Australasia.

The tree ferns of the Carboniferous period played a crucial role in carbon sequestration. As these plants photosynthesised (generated energy) and grew, they captured atmospheric CO_2, storing it in their tissues. Over time, as the tree ferns died and accumulated in swampy environments, their organic matter partially decomposed, leading to peat formation. Under high pressure and heat, this peat eventually transformed into coal, which became an abundant energy resource millions of years later.[5] A few years ago, I visited a forest in Australia known as the Cathedral of Ferns. This tells you something about their majesty and size. They grow at around 25 mm per year, and the towering specimens in the Cathedral of Ferns were up to 15 m tall, so potentially around 600 years old. Tree ferns are also hardy. In the aftermath of a forest fire, amid the sombre ashy backdrop, the unfurling of the jade-green fiddleheads is a potent sign of regeneration.

Zoom forward in your time machine to around 330 million years ago and head to Scotland. You might now see a tree called *Pitys withamii*. The landmass that's now Scotland was once part of a tropical swampy forest on the supercontinent of Pangaea. I learnt about this tree after visiting the Natural History Museum in London. Adjacent to the Palaeontology wing, a colossal fossil monolith stands proudly outside. The fossil is a petrified *Pitys withamii* brought to London after being discovered in a quarry in Scotland over 130 years ago. The fossil, weighing around 11 tonnes, is so heavy owing to fossilisation, where minerals replace organic material. Indeed, during the Carboniferous period, the tree would have been three times lighter than it currently is, but still a mammoth plant of its time.

A giant Sequoia redwood tree. Note the horses in the bottom left of the image for size reference. This tree was not growing in its native environment but in Hampshire, England (likely brought over here by the Victorians).

Some of the coniferous trees in the Carboniferous period were the ancestors of the great Sequoia trees (see image above) that tower above all other plants today. Sequoias are long-living and among the tallest trees in the world. Their branches and needles provide habitat and nourishment for various creatures, while their presence alone can inspire awe and reverence. A cross-sectional piece of a giant Sequoia tree wows the visitors of the Natural History Museum in London. It was 1,300 years old and 101 m tall when it was felled in 1891. No doubt it would have lived for much longer back when humans and their giant chainsaws weren't even a twinkle in the universe's eye.

The earliest trees on the supercontinent of Pangaea were a group called *Archaeopteris*, which stood around 10 m tall, about the height of a mature silver birch tree today. However, unlike most modern conifer trees (called gymnosperms from the Greek gymnos for 'naked' and sperma for 'seed'), which reproduce via seeds, the *Archaeopteris* reproduced via spores, much like ferns do today.[6]

In the late Carboniferous period, seed-bearing plants started to evolve. Plants in the genus *Cordaites* were one of the largest trees in this period, reaching 30 m with branches shooting out from the stem.[7] As the climate shifted towards the end of the Carboniferous period, the early pioneers gave way to gymnosperms, and conifer trees began to flourish.

It's time to turn the dial again. This time you're heading to the Triassic period (248 to 206 million years ago). Primitive conifers continued to dominate, with many trees resembling the cycads and monkey-puzzle trees we see today.[8] These gymnosperms dominated the northern landscapes when flowering plants and grasses hadn't yet evolved. Animals and plants were recovering from the great Permian mass extinction event, which wiped out 83% of marine species and 70% of terrestrial vertebrates – the most severe mass extinction in history.[9] The understorey of these primeval woodlands would have been thronging with other species that no longer exist, including many mosses, ferns, shrubs and woody vines. The forests also supported plant-eating dinosaurs, which evolved during the Triassic period.[10] Some of these were specialist eaters of low-lying plants, while others could reach up and browse the taller forest vegetation, much like the giraffes on the African savannah today. Beetles and scorpionflies would have lived in and around the forests, and the soil and plants were heaving with invisible biodiversity (bacteria, algae, viruses and others). This vast array of creatures contributed to the ever-evolving web of life in these ancient forest landscapes.

The break-up of Pangaea

Towards the end of the Triassic period, the supercontinent of Pangaea, which incorporated almost all the landmasses on Earth at the time, started to break apart.[11] The mechanism for this break-up is now explained by plate tectonics and the immense geological forces that pull continents apart. Following this break-up, two large continents were formed – the northern continent Laurasia and the southern continent Gondwanaland. At the time, the two supercontinents displayed distinct botanical profiles. Conifers and other seed plants dominated Laurasia, whereas seed ferns dominated Gondwanaland.

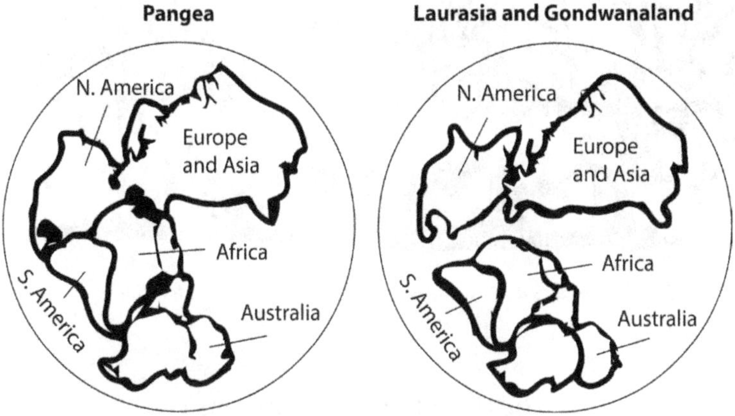

The break-up of Pangea formed two supercontinents – Laurasia (top) and Gondwanaland (bottom).

If you turn the dial so it points at 206 to 144 million years ago, you'll arrive in the famous Jurassic period. Considerable changes have occurred in animal populations. Enormous sauropods such as the Diplodocus and Brachiosaurus devoured vast swathes of coniferous foliage. However, their browsing strategy was probably diverse, eating ground vegetation and canopy-level foliage.[12] With their long necks and powerful jaws, they effortlessly plucked mouthfuls of foliage, consuming copious amounts of plant material in a single feeding session. These huge creatures likely shaped the forest landscapes by trampling forest vegetation and creating vast expanses of savannah-like open spaces.[13] They were ecosystem engineers, moulding this ancient land. They probably contributed to the selective pressures that shaped the evolutionary adaptations of both plants and herbivorous dinosaurs alike.

The climate during the Jurassic period was mostly warm and moist, the sea levels were high and there's no convincing evidence of polar ice.[14] Temperate and subtropical forests were dominant, and towards the end of the Jurassic period a group of new kids were on the block – the flowering plants or 'angiosperms'.

Monkey puzzles and cycads were important vegetation for dinosaurs during the Jurassic period.

The rise of the angiosperms

Flowering plants or 'angiosperms', from the Greek *angeion* for 'vessel' and *sperma* for 'seed', evolved around 130 million years ago.[15] They quickly spread across the world and dramatically changed the Earth's landscapes. Angiosperms are fast-growing and highly adaptable, which led them to take over many ecological niches towards the end of the Cretaceous period. The plentiful and nutritious food source of these flowering plants likely led to a boom in dinosaur populations during the Late Jurassic and early Cretaceous periods.

Some flowering vegetation in and around Cretaceous forests included magnolias, early sycamores, palms, laurels and barberry.[16] Cretaceous vegetation increased dramatically in diversity and density, and the rapidly adapting angiosperms radiated across the planet, leading to diverse forest types. In my mind, I picture this evolutionary radiation event as a fireworks display. Much as fireworks explode into various colours and patterns, evolutionary radiation diversifies single ancestral species into multiple descendant species with unique traits and adaptations. Evolutionary radiation can also occur at different rates and with varying levels of impact. Some groups of organisms may experience rapid diversification, while others may evolve more slowly over longer periods.

Speaking of magnolias, I recently chatted with my friend Harry Watkins, the director of St Andrews Botanic Gardens in Scotland. He told me that the Cretaceous period magnolias were one of the first documented plants to utilise pollen with fragrant scents to attract pollinators such as beetles and flies. They were one of the dominant plant species of the time, growing in diverse environments ranging from tropical rainforests to swamps and providing food and habitat for various organisms, from insects to dinosaurs. The magnolia forests would have been bustling with activity, with animals of all shapes and sizes going about their daily lives in a lush and vibrant ecosystem.

Forests during the Cretaceous period were highly productive. Indeed, tree-ring data suggest that annual forest growth was considerably higher than today, particularly in areas at mid- to high latitudes. The increased productivity of Cretaceous forests is attributed to fertilisation by elevated CO_2 levels and the expansion of tropical and temperate belts. Towards the end of the Cretaceous period, angiosperms were the dominant vegetation in forests.

Boom!

It would be unadvisable to turn the dial to 66 million years ago – the time of the K-T event, sometimes called the K-Pg event or Cretaceous-Paleogene event. Another mass extinction occurred, and around three-quarters of the animal and plant species on Earth were wiped out.[17]

The leading theory is that an asteroid 10 km wide hit the planet in the Gulf of Mexico region, although, as with many theories, the extent of its impact is still up for debate. The most likely site of the crater, which still exists today, is called Chicxulub. The impact event was an approximately 100 million-megaton blast generating a core of superheated plasma. Some theories suggest that the dust and sulphate aerosols injected into the atmosphere after the asteroid impact caused a darkening of the skies as if a celestial deity had dropped an Earth-sized blanket over the heavens. Because plants depend on light to generate energy, this blanket in the sky would have caused dramatic changes to plant communities worldwide. But some of the original plants were resilient and adaptable. They soon (evolutionarily speaking) recovered, and continued to diversify in the new climate of the Tertiary period, 66 to 2.6 million years ago.

Flowering plants remain the dominant plant group on Earth today: around 80% of all plant species are angiosperms. However, grasses evolved in the early Tertiary period, leading to richer mosaics of forests and meadow-like habitats, spreading out like a vibrant tapestry across the land.

You'll need to take plenty of water to hydrate on your next journey to 55 million years ago (the Eocene). The global climate underwent extreme warming, probably because of large amounts of methane being released from the ocean floor. Consequently, the warming climate led to trees growing even in polar regions, while subtropical or tropical regions were covered in angiosperm forests. Palm trees even grew in Alaska![18] Eucalyptus trees evolved in Gondwanaland when Australia was still attached to Antarctica. The eucalypts adapted to the drier climates and fire-prone landscapes as Australia drifted north.

Cooling

Global cooling began during the Oligocene period (33.9 to 23 million years ago). This later led to widespread glaciation in the northern hemisphere. Forests began to shrink, grasslands expanded and mammals diversified. The late Tertiary forests of the land we now call Europe were dominated by conifers. But deciduous trees were dominant here by the Quaternary period (beginning 2.58 million years ago and when our human ancestors first evolved).[19] The continental climate was less favourable to the conifers. Harry told me that across the pond in North America, the temperate deciduous forests are similar to those in the late Tertiary period. Spruce forests grew in the Great Plains and pine forests dominated the South Atlantic coastal plain.

The colder periods of the Quaternary ice age caused glaciers to cover vast areas of the globe, and trees failed to grow. But during the interglacial periods (intermittent warming), plant life in these areas would flourish again. It's quite astonishing to think that ice covered 25% of the Earth's land during the last ice age, and ocean levels dropped to 125 m lower than today.[20] This meant that land area in some places of the planet increased. Previously hidden landmasses the size of South America appeared, and Australia, for instance, was 20% bigger than it is now.[21] Therefore, despite vast areas of forests disappearing in the northern hemisphere, it's likely that new forest ecosystems grew in these previously submerged landmasses, which are now, once again, hidden deep under the sea.

You'd be forgiven for assuming forests in Europe were dark and dense during the last Interglacial period (129,000 to 116,000 years ago). However, new evidence suggests that light and open woodland represented over 50% cover in temperate regions during this period–possibly maintained by large herbivores.[22]

There'll be plenty of fellow humans walking around when you turn the dial to 11,700 years ago. The last glacial period concluded, leading us into the present interglacial period, characterised by around 10% ice cover, mostly at the poles. During this warmer phase, forests experienced a resurgence, yet there was an unabated and rapid rate of deforestation.

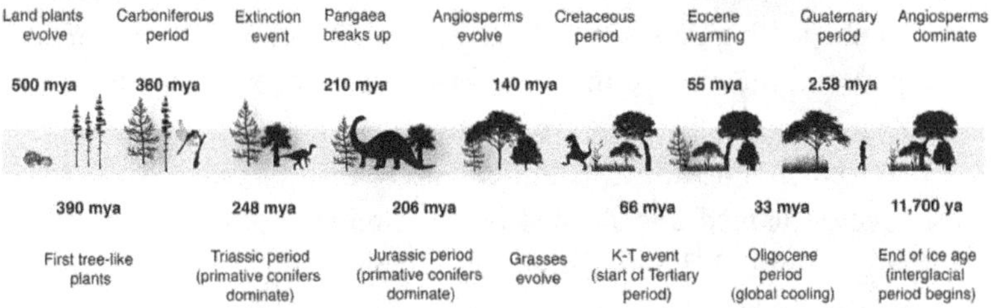

Timeline of forest evolution.

Wildwood

There's a long-standing debate in academia about the concept of 'wildwood'. This term was popularised by the late historian and ecologist Oliver Rackham. It describes the state of woodlands, particularly in temperate climates, before humans manipulated them. One of Rackham's most influential books was *The History of the Countryside*,[23] in which he explores landscape features: fields, roads, hedgerows, fens, marshes, rivers, coasts, heaths and woodlands and their evolution and development over the centuries. It describes the landscape with its many contours and textures as if the landscape were a book itself: the lumps and bumps, the colours and the shape of the trees tell a story of the landscape's history and its interactions with humans.

I know of people who went for walks with Rackham. They'd take him to a new place, somewhere he'd never been, and he'd instantly start reading the landscape. He'd talk deeply about the structure or composition of the trees and the forest. And then suddenly there'd be silence. You'd look around and he'd be taking notes. Later, weird and wonderful stories of lichens would liven up lunch.

Rackham once said that "trees are wildlife, just as deer or primroses are wildlife. Each species has its own agenda. And its own interactions with human activities. That is, we should remember that each wood is different to every other wood. A plantation is not a wood. And mass tree planting is not conservation. It is proof that conservation has failed." His writings profoundly changed the way I viewed the countryside.

As a young child, when I visited the Peak District National Park in Derbyshire, England, I marvelled at the open moorland with scattered blocks of trees in the background. Who was I to question how 'natural' this was? I probably assumed these were untouched natural habitats that had evolved over thousands of years, whereas human-shaped habitats were places such as towns and cities, farms and parks. Little did young Jake know that the moorlands were heavily managed (burnt) for grouse shooting, and the uniform blocks of trees were monoculture plantations – essentially, a group of a single tree species planted and farmed by the timber industry. Therefore, all these 'natural' habitats I saw were highly shaped and controlled *by* humans *for* humans. But even in less-managed landscapes, there were signs of historical interactions between humans and the forests that some may consider 'wild'. For example, large earth mounds in the forests of Europe may seem like natural variations in topography, but they're often the remains of prehistoric burial mounds dating back to the Bronze Age. You may find wood banks with ditches and veteran trees on top. Humans often created these to mark the boundaries of their parishes. The types of trees and how they grow also tell a story of human interaction. Sometimes, they are ancient coppice stools with trees cut close to the base to encourage subsequent growth. Bumpy hollows in the ground are often the remains of saw pits used by woodland workers in bygone eras.

Shifting baseline syndrome

There is a proposed phenomenon known as shifting baseline syndrome or environmental generational amnesia. Simply put, this is a gradual change in the accepted norms for the condition of the natural environment owing to a lack of experience, memory and/or knowledge of its past condition. Soga and Gaston (2018) argue that without memory, knowledge or experience of past environmental conditions, people born now (i.e., current generations) find it harder to perceive how much their environment has changed.[24] This is because they compare it to their own 'normal' baseline and not to the baselines of the past. Shifting baseline syndrome can be detrimental because it can increase our tolerance for progressive environmental degradation, wildlife population declines, increased pollution and habitat loss. For example, a person who grew up near a polluted river may view the state of that river as normal, while a previous generation may remember it as clean and pristine. Over time, each generation may become more accepting of degraded environmental conditions, resulting in a 'shifting baseline' that fails to accurately reflect the true state of the environment and the extent to which it has altered.

I wonder if the spiritual dangers mentioned in the Introduction, of supposing that food comes from the grocery and paper from the stationery shop, are exacerbated by – or are a symptom of – shifting baseline syndrome. Suppose we're privileged to be born into convenience at a click without being nurtured to respect the source of our commodities. In that case, our normal baseline is frighteningly depleted of those deep organic bonds that guide us to protect the natural world.

Soga and Gaston use Japanese forests as a case study for shifting baseline syndrome. They say that Japan is one of the world's most forested countries, with over 70% of its land covered by forest. People born five or ten generations ago would see primary (ancient, biodiverse) forests everywhere. However, modified forests (e.g., monoculture plantations for timber) have gradually expanded across the landscape with each passing generation. By the year 2000, most primary forests had been destroyed and replaced by plantations, thereby establishing a new baseline. Current generations now see these plantations as how nature is supposed to be – it's their norm.

We often rely on written accounts, signs in the landscape and radiocarbon analysis to understand what the landscape was like in the past. Sometimes these data are a bit ropey, and it becomes difficult to paint an accurate picture of historical baselines. Going back to the concept of wildwood – the woodlands that existed before humans started to heavily manipulate the landscape – well, there are some debates about what it looked like, particularly in the temperate climate of the northern hemisphere. There are no written records of what woodlands looked like 11,000 years ago. The forests we see today are a legacy, but a very altered legacy of the wildwood. During the last ice age, much of Britain was covered in ice. Therefore, trees were unable to grow. At the end of the last ice age (around 11,000 years ago), the post-glacial warming climate and the availability of land allowed trees to re-establish and disperse. Birch and Scots pine were likely among the first pioneering species to establish, followed by hazel, elm, oak and others.

The traditional view is that wildwood was a continuous climax forest with an almost entirely closed canopy covering the entire country of Britain. I remember one of my old tutors, Dr Rowan, joking, "Some say a squirrel could hop over the treetops from Land's End to John O'Groats without touching the ground." Rackham thought this was inaccurate, but that wildwood landscape was still largely a closed canopy interspersed with small but significant gaps.

Cue the Vera hypothesis. Dutch ecologist Frans Vera thinks we've got it all wrong.[25] He says the closed-canopy hypothesis is inaccurate because we've not paid enough attention to vast herds of grazing herbivores. When large herbivores feast on the vegetation, they create a mosaic of open land with interspersed forest patches. This hypothesis gave rise to enormous controversy and arguments in the field. It helps explains some of the things that have been a problem for the old closed-canopy hypothesis. Still, it rests on assumptions about there being a considerable number of animals to produce this pattern of vegetation dynamics on a large scale. Unfortunately, we can't prove either of these things.

We can turn to pollen grain evidence, indicating the vegetation in the landscape across different ages. Radiocarbon dating allows us to map how forests have grown and evolved over time. If you look at the pollen record for Britain, there's an inordinate amount of oak and hazel pollen. These trees regenerate optimally in open environments; that is, they don't like shade and prefer to grow in thorny thickets where animals can't graze them. Vera postulates that if these species are regenerating prolifically, as the records indicate, large herbivores must have maintained vast areas of open habitats.

However, using the island of Ireland as a case study, other researchers have rejected Vera's claims. At the end of the last glaciation (11,000 years ago), when trees moved into Britain, the English Channel was dry. Therefore, trees could migrate on animals and the wind via the land bridge connecting England to mainland Europe. As such, large grazing animals could migrate to England. However, Ireland was isolated long before the English Channel was filled with water. This meant that large grazing animals could not naturally reach the island, and the woodland in Ireland must have developed over thousands of years without vast herds of these large herbivores. Importantly, researchers have found no significant difference between the pollen records of the island of Ireland and those of England; that is, hazel and oak were present in Ireland in similar quantities to England. So the assumptions of Vera's hypothesis are questionable.

Vera points out that oaks are failing to reproduce in existing shaded woodlands right across Europe today. He thinks the fault lies with conservation biologists and foresters who have kept grazing herbivores out of the woods for fear of browsing damage. He believes that the spreading canopy of darkness has prevented savannahs from remaining open by interrupting the reproductive dynamics of the vegetation. Rackham disagreed with this perspective. He said that oaks were successfully reproducing in existing woodlands until the arrival of American oak mildew in the early twentieth century. This, Rackham claimed, had the effect of inhibiting photosynthesis and forcing oaks to require more light than they had historically received.

This major toing and froing between viewpoints continues today, reflecting how little we know about historic landscapes. The structure of British wildwood was probably intermediate between these extremes – the continuous closed canopy and the vast areas of open land. It's also possible that we fail to consider the contribution of human activities towards maintaining large open spaces, even in early prehistory. Indeed, it would have been advantageous for Mesolithic hunter-gatherers to augment and manipulate open grassy areas where wild grazing animals gathered and encourage nuts and berries to grow along woodland edges.

The Biggest Estate on Earth by Bill Gammage and *Dark Emu* by Bruce Pascoe are two myth-busting books.[26] They counter the notion that Australia was an untamed wilderness before Europeans arrived, and reveal a complex, country-wide land management system applied by the Indigenous Peoples of Australia. Personally, I think it's telling that many of us find this surprising.

For thousands of years, Indigenous Peoples set fire to and managed the land far more systematically and scientifically than most people realised. Across Australia, early European settlers repeatedly commented that the land looked like a country park with expansive grassy areas, pathways and open woodlands with diverse and abundant wildlife. Therefore, our outdated vision that forests are a continuous stretch of unaltered and untamed closed-canopy woodland turns out to be a myth in more than one place in the world.

However, many Indigenous Peoples across the planet are known to have coexisted harmoniously with their forests for millennia while conserving the forest's biodiversity. A recent study into the history of the Amazon rainforest found that some Indigenous Peoples lived there for thousands of years, "causing no detectable species losses or disturbances".[27] Now the Amazon and other forests where Indigenous Peoples still live are being deforested at alarming rates. Some reports suggest that a football pitch-sized area (7,140 m^2) of tropical rainforest is lost every six seconds.[28]

A note on kelp forests

We don't usually talk about underwater communities as being forests. And kelp isn't technically vegetation or a plant per se – it's giant algae. Kelp forests do, however, share some parallel traits with terrestrial forests. Much as in a dense forest on land, losing oneself among the giant tangled fronds and blades of a kelp forest would be easy. Like trees in a forest, kelp forests provide food and shelter for many creatures. They harbour a higher diversity of plants, animals and microbes than many other ocean habitats. They're a magical world beneath the water's surface, where sunlight streams through kelp fronds like golden glades. Sea otters frolic in the kelp, rolling and tumbling, while sea urchins and abalone cling tight to the swaying sea. If you were lucky to dive into a kelp forest, you'd probably see colourful fish darting in and out of the kelp, like rainbows in the sea; seals and sea lions also glide gracefully through the fronds in search of food.

The kelp forests require light to photosynthesise and experience seasonal changes, much as temperate land forests do. They prefer nutrient-rich waters around the world, primarily in temperate and polar regions, and cover around 25% of the Earth's coastlines – an estimated area of over 2 million km^2 or about the size of the Democratic

Republic of Congo.[29] Kelp species evolved millions of years ago but are thought to have peaked between 20,000 and 7,500 years ago as rising sea levels created new habitats. They subsequently shrank by an enormous 70% to today's levels. Humans have harvested kelp as far back as the Neolithic period (10000–4500 BCE), possibly further, and continue to do so today. This underscores that interactions between forests and human beings, whether on land or at sea, have been taking place for countless centuries.

It's important to understand the history and evolution of the trees and forests that once lived and those that still grace our lives today. But knowing how, when and why we're losing them is also vital. Otherwise, we run the risk of sleepwalking into an existential crisis.

CHAPTER 2

A Brief History of Deforestation

"Silencing of the rainforests is a double deforestation, not only of trees but a deforestation of the mind's music, medicine and knowledge."
—Jay Griffiths

Deforestation – the purposeful clearing of forested land. Is this a modern-day issue? Certainly, but it's also deeply rooted (pardon the pun) in thousands of years of history. People have been chopping forests down for millennia. Some cultures have contributed to deforestation more than others, an unnerving disparity that continues today. But how much of the world's forests have we lost? How did this happen, and why? Read on to find out – but be warned: a chapter on deforestation is fascinating but can in places be quite heart-rending by its very nature. However, I promise that stories of connectedness and hope will arrive soon.

Around 10,000 years ago, nearly 60% of the world's habitable land was covered by forests of various densities.[1] This was approximately 6 billion ha of land. For reference, take a moment to think about the size of Russia, the largest country in the world (1.71 billion ha), and then expand that image by three and a half times. That's an enormous landmass covered in vast forest ecosystems. The rest of the Earth's terrestrial surface was primarily wild grassland, shrubs and biocrusts, interspersed with freshwater habitats. Humans had been felling trees to build shelters and make tools for centuries with minimal impact on the integrity of forest ecosystems. However, the dawn of agriculture, which arguably occurred at different times across the planet, dramatically accelerated the rate of deforestation as humans cut down trees for permanent dwellings and to clear the land for livestock and crops. While half of this deforestation occurred between 10,000 years ago and the turn of the twentieth century (only 10% was lost between 10,000 and 5,000 years ago), the other half – around 1.1 billion ha, or the combined area of the USA and Greenland – has been lost since 1900. A large portion of this has been depleted in the last couple of decades, and currently, over 10 million ha of forest are lost each year.[2]

If we view this all through a Darwinian evolutionary lens, an intriguing question arises: how might the traits and behaviours that historically benefited human survival and reproduction also play a role in the phenomenon of deforestation?

Throughout our evolutionary history, the acquiring and use of resources has been crucial for survival. Expanding into forested areas and utilising their resources, such as timber for shelter, fuelwood for warmth and cooking, and land for agriculture, provided an advantage in terms of securing food, shelter and energy. Humans have displayed remarkable adaptability, which has allowed us to thrive in various environments. This adaptability also includes being able to modify landscapes to suit our needs. Clearing forests and transforming them into productive environments for agriculture, settlements and infrastructure has helped our species expand. Our evolution is closely tied to our ability to innovate. Throughout history, advancements in tools and technologies have increased our capacity to modify and exploit natural resources, including forests. From early axes to modern machinery, technology has enhanced the efficiency and ubiquity of deforestation.

It's also important to note that human behaviour is influenced by a multitude of factors beyond genetics. Indeed, our traits are shaped by a myriad of cultural and environmental factors. For instance, in cultures where trees are revered as sacred, there may be traditions of forest conservation and rituals honouring tree protection. Conversely, in societies where trees are primarily viewed as economic resources, deforestation for timber extraction or agricultural expansion may be more prevalent. Our understanding of the consequences of deforestation and its long-term impacts on planetary systems and processes has also evolved. As we gain a deeper understanding of the interdependence of ecosystem health and sustainable living, which seemingly evolved faster (or at least lasted longer) in some cultures than others, there's been a surge in efforts to protect nature and curb deforestation.

Deforestation and the dawn of agriculture

It's often said that the dawn of agriculture was the main driver of prehistoric deforestation. But when and where did it all begin? Over 10,000 years ago, eight Neolithic founder crops were cultivated in the Levant, a large area of the Eastern Mediterranean region of Western Asia.[3] These crops included chickpeas, lentils, peas, wheat, barley, vetch and flax. This is some of the earliest known arable activity on Earth. However, pigs, sheep and cattle were domesticated in a similar area and likely before the Neolithic crops were mainstream in the region. Growing evidence for early agriculture in other areas of the world also exists. For instance, several plant species in South America were cultivated over 8,000 years ago. The humble potato was first domesticated in the Andes region between 8000 and 5000 BCE, followed by beans, tomatoes and peanuts.[4] It's likely that trees made way for such crops. However, the extent of the impact on forested landscapes was considerably less than today.

In the Amazon, at least 9,000 years ago, humans found a plant known as cassava. I can picture the early cultivators carefully selecting and nurturing the most promising plants. They would have observed the different varieties, tested various cultivation techniques and shared their knowledge from generation to generation. It was a pillar of the annual crops and became a staple food of the native populations in South America.[5] I recently saw cassava grown in among forest trees on a visit to the Fiji Islands. It's hard to believe that mass forest clearance was needed to make way for such crops (particularly if these mixed forest-crop systems were used), but it likely contributed to deforestation on a small scale.

In New Guinea, sugarcane and various root vegetables were also domesticated in around 7000 BCE. Still, the dominant historical narrative has characterised the Indigenous Peoples of this area and further south in Australia as almost exclusively nomadic hunter-gatherers. This may be the case for some peoples; yet as decolonisation processes gain traction, there's growing evidence that Indigenous Australians cultivated crops thousands of years ago to make bread and other foods using kangaroo grass and yams.[6] They also used sophisticated eel traps over 8,000 years ago.[7] There's evidence that forests were managed by Indigenous Peoples in pre-1778 Australia to enable crop production, and controlled burning was often the tool of choice.[8]

The cassava plant has bulbous and highly nutritious roots.

In Asia, rice was cultivated at least 8,000 years ago.[9] It is, however, unlikely that these agricultural practices significantly impacted forested landscapes in the early days. Traditional methods such as shifting cultivation or slash-and-burn techniques were commonly employed in the early stages of rice cultivation. These methods involved clearing small forest patches to make space for rice paddies. Over time, as the population increased, so did the impact on forests.[10] However, the cleared areas were often left fallow for extended periods, allowing natural forest regeneration to occur.

The last few millennia

"We'll go camping in a never-ending forest," remarked Lukáš, a Slovakian friend. He was excited to take me on a two-day trip to a dense forest where families of bears, wolves and wild boars roamed freely, and where you could walk for days with only the trees and more-than-human forest dwellers for companions. I was a little sceptical, having grown up in England, where forests were sparse and megafauna was minimal. However, upon hearing more of Lukáš's stories and researching the trip, my excitement grew exponentially. I arrived in Slovakia, and after a night of beer, guitar playing and catching up, we headed into the forest. It was like stepping into a fairy tale, complete with giant creaking trees, waterfalls, sounds from elusive creatures and hidden wonders.

One of the two nights, we decided to sleep in our bivvy bags outside the tent to watch the stars. At one moment, when Lukáš was fast asleep, I remember hearing several twigs snapping along with snorting and snuffling sounds getting closer and closer to me: a family of wild boars charging their way through the forest. I caught a glimpse of at least four boars lit up by the moonlight, one scurrying over the bottom of my sleeping bag. My heart was pounding like a wild drum, echoing the exhilaration that coursed through my veins. This experience made me wonder what the forests – with their wonderful wildlife – must have been like in the rest of Europe before the surge in deforestation. Today, Slovakia has over 40% forest cover, but around 2,000 years ago as much as a startling 80% of Western Europe was covered in forests.[11] Then agriculture rapidly spread across Europe, and today the average forest cover is just 33%.[12] Deforestation during the Roman period (27 BCE to 476 CE) was prolific as the territories of the Empire expanded across Europe. However, there's limited evidence of direct Roman military campaigns or administrative control in the area that is now Slovakia, which may have helped preserve the forests there.

With the Roman expansion came a boom in population growth, large-scale agriculture and rapid urban development. This resulted in a barrage of deforestation across Europe, particularly in the Mediterranean region. However, there's a myth that Britain was a 'frontier province with boundless woodlands' until the Romans arrived. As discussed previously, some academics think 50% of the forests in Britain were lost by the Iron Age (2500 BP). The scribes of the Domesday Book (published in 1086) indicated that around 15% of English land was woodland, which is close to today's proportion – and a far cry from that in Lukáš's homeland. Forest cover is a mere 10% in England today, and much of that is monoculture plantation.

The Nazca

Despite studies showing that Indigenous Peoples in the Amazon coexisted with forests for millennia and caused no detectable species losses or disturbances, some evidence suggests that deforestation did exacerbate extreme climatic events, leading to the collapse of some societies. Take the Nazca civilisation, which flourished in the arid coastal region of Peru between 100 BCE and 800 CE. The Nazca society was made up of several chiefdoms and developed sophisticated agricultural techniques, including substantial aqueduct systems for irrigation called *puquios*. These spiral features funnelled into the ground to help draw wind down to the underground canals, moving the water through the system.

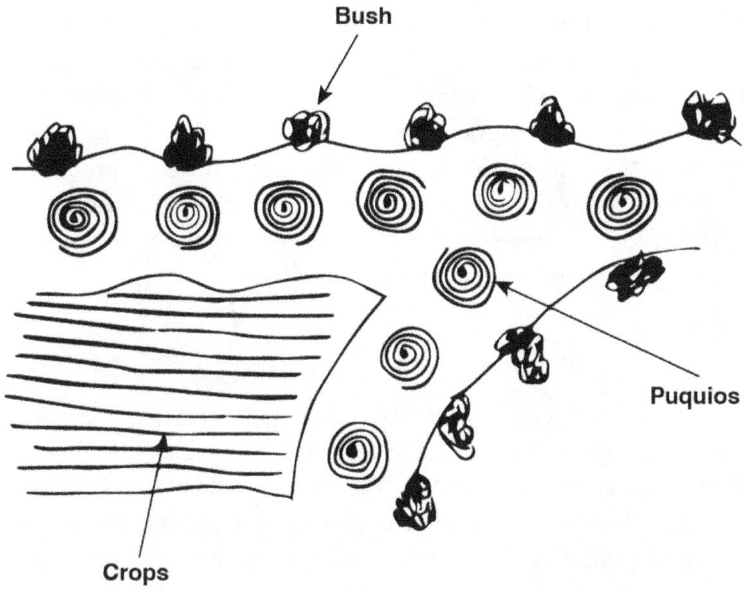

A bird's-eye schematic of the Nazca puquios *hydrology system.*

However, from around 500 CE, the civilisation began to decline, and by 800 CE, the civilisation had completely collapsed.[13] The primary reason for this is thought to be unusually harsh El Niño events (warming of surface waters in the eastern tropical Pacific Ocean) that triggered droughts and destructive flooding. However, evidence suggests that these impacts were exacerbated by deforestation – specifically, the felling of *Prosopis pallida* trees to make way for crops.[14] The trees were legumes that could fix nitrogen in the soil, improve the soil quality and, in symbiosis with a gluey protein called glomalin, strengthen the soil structure. This meant the trees were vital in preventing flooding and wind erosion. So, it's likely that a combination of floods, droughts and soil erosion (by removing trees) put immense pressure on the Nazca society.

China

In China, agriculture had been practised for thousands of years before the Shang dynasty was established in 1766 BCE. With agriculture came deforestation. The Shang dynasty was one of the earliest civilisations to document its activities with written evidence, so there are additional clues. It was during these times that urban areas expanded dramatically. This required land clearance for space and materials for buildings. The Zhou dynasty overthrew the Shang dynasty in around 1150 BCE, and by the time the Zhou dynasty peaked in 500 BCE the population of China had exploded – increasing from 5 million (during Shang times) to 30 million. The Chinese population explosion came with increased agricultural and infrastructural demands, and improvements in technology intensified. Large furnaces producing swathes of valuable bronze items required temperatures of 1,400–1,535°C. These furnaces must have demanded a considerable amount of wood and, therefore, the mass felling of trees.

Africa

There's evidence of significant forest loss in Africa between 2,000 and 3,000 years ago. At the time, Neolithic farmers expanded southward from Nigeria and Cameroon. There's a question whether these Neolithic farmers took advantage of a retreating forest owing to a changing climate or whether the farmers were primarily responsible for the retreating forest. Recent evidence suggests no significant climate change during this period, implying that the forest loss might have been caused by farm expansion.[15] Archaeological evidence shows that iron metallurgy and pearl millet farming were rife in this area at the time.[15] The demand for iron ore meant that vast amounts of charcoal were needed and land clearance expanded to meet agricultural needs. In any case, the forests certainly retreated during this period, but whether the retreat was caused by the climate, human activity or both is still up for debate.

The collapse on Easter Island

There's a famous story about deforestation on Easter Island in the south-eastern Pacific Ocean. According to radiocarbon dating, the island was settled by Polynesians in around 800–1200 CE.[16] It's often said that Easter Island (Rapa Nui) is one of the most extreme examples of deforestation globally and that this led to ecological and societal collapse. Indeed, the island was covered in palm trees for over 30,000 years, but the entire forest is now gone and all native tree species are extinct. The traditional narrative posits that the Polynesian islanders started to harvest the forest unsustainably in around 1200 CE, peaking in 1400 CE. And by the time the Dutch explorers 'discovered' Easter Island in 1722, the whole island was completely deforested.[17] However, the idea of an abrupt island-wide event said to be caused by the islanders has recently been challenged. Recent sediment core analysis has revealed that deforestation was not abrupt but gradual, and its reasons are more complex than initially thought: droughts and other ecological factors contributed to the forest's demise.[18] Archaeological evidence also suggests that the Polynesian islanders were resilient to deforestation and remained healthy until Dutch contact, contradicting the notion of self-induced societal collapse. In any case, it's clear that Easter Island's forests were wiped out and humans played a role.

Colonialism

Between 5,000 years ago and the start of the eighteenth century, the global population increased more than tenfold to over 600 million people.[19] Agricultural land was expanding dramatically, and many trees were needed to fuel Western colonialism and the many European naval wars. The footprint of woodlands required to construct the British fleet alone – which in 1812 consisted of 130 large battleships with 60–120 guns and a further 600 frigates and smaller vessels – must have been astronomical. HMS *Victory*, for instance, was built of more than 5,500 oak trees.

Following the arrival of European settlers in the USA, vast swathes of the forests in the eastern part of the country were felled between 1620 and 1900 for timber and agriculture. I've heard people say things like: "It was an era where the American dream beckoned, and opportunities seemed boundless." The nation was swept up in a whirlwind of innovation. People flocked to build settlements, seeking new beginnings and the promise of a brighter future. But alongside this urban growth, a remarkable landscape transformation was taking place. Extensive stretches of virgin forest were eradicated between

The restored HMS Victory *in Portsmouth's dockyard is a potent reminder of deforestation in the name of defence and colonialism.*

1850 and 1926 (see the maps based on the original work of William B. Greeley – chief of the US Forest Service from 1920 to 1928).[21] From one perspective, the maps could be misleading as forest restoration has been considerable in the east of the country during the last century. However, the maps are still important for visualising the scale of historical deforestation of virgin forests.

During the nineteenth century, the population of the USA increased by nearly 15 times (from around 5 million in 1800 to 75 million in 1900).[22] The total land area converted to agricultural fields (for crops and grazing) increased by over ten times in that century, from 26 million to 298 million ha.[23] So, for every person living in 1900 in the USA, around 4 ha of land were ploughed for crops or grazing.

As I've discussed in another book (*Invisible Friends: How Microbes Shape Our Lives and the World Around Us*), this extensive ploughing of the land eventually led to a vast area becoming eroded in the early twentieth century. The area was known as the Dust Bowl, and thick dust clouds travelled hundreds of miles, coating everything in their path – storms being known as black blizzards.

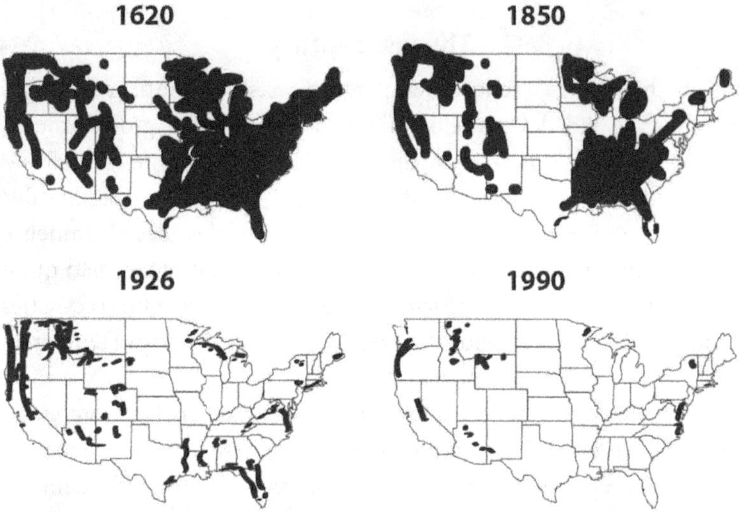

Dark areas show the presence of virgin forest between 1620 and 1990 in the USA.

The key statistics here – that 50% of global deforestation occurred between 10,000 years ago and 1900, and the other 50% occurred in the last century – highlight two vital points. First, deforestation is by no means a novel problem. Relatively small human populations of the past were capable of chopping down forests at quite a rapid pace. Even with reasonably basic lifestyles (from a technological and consumption perspective) compared with today, the per capita footprint of our ancestors would have been significant in many areas of the world. Agricultural productivity was also relatively poor by today's standards (although nutrient content was likely higher owing to soil quality). This, along with the reliance on wood for fuel, meant that swathes of land were cleared for fundamental provisions. Secondly, it emphasises how much deforestation has increased in speed over the last century. If we think about the history and evolution of forests, 100 years is a drop in the ocean. Yet in the last 100 years, the world has lost as much forest as it had in the previous 9,000 years.

Let's take a moment to think about this land clearance. You may be picturing urban sprawl covering the land where dense forests once anchored their roots. However, 'built-up' urban land only accounts for less than 1% of our planet's habitable land.[24] Our biggest footprint comes from what we eat, not where we sleep. Indeed, most forest clearance is to make way for livestock and crops.

The last century

Before the twentieth century, temperate forests were the primary target for logging and deforestation. Dramatic population growth meant the demand for resources in the so-called high-income countries – for agriculture, energy and infrastructure – increased dramatically. The burgeoning populations in Western countries placed immense pressure on their domestic resources, leading to an intensified quest for new territories and raw materials. Expanding agricultural activities required vast tracts of land, often resulting in the clearing of temperate forests to make way for farmland.

The accelerated demand for agricultural land, energy from wood and timber for infrastructure in the twentieth century increased with technological advances and urbanisation. With this demand came further deforestation. During the two world wars (1914–1918 and 1939–1945), trees were needed for fuel, ammunition crates, carts, wagons and fortifications. During the First World War, Europe experienced destruction on an unprecedented scale. Millions of tonnes of explosives bombed entire forests out of existence. These precious ecosystems, which support myriad species, are the untold victims of human wars. They're destroyed to fuel destruction and then destroyed some more – the worst kind of vicious cycle. And the cycle is completed by reducing their loss to 'collateral damage'.

Between 1920 and 1980, forest losses quadrupled to 120 million ha per decade.[25] That's equivalent to the area of South Africa. Europe and the USA were now importing considerable timber produced by felling trees in the tropics. And when the economic expansion of countries across Asia and Latin America was added to the mix, global forests were under colossal pressure.

Global deforestation reached its peak in the 1980s. During this decade, 150 million ha of forests were lost.[26] That's equivalent to an area nearly twice the size of Pakistan in just one decade. Clearing the largest rainforest on Earth (the Amazon) for croplands was a significant driver of this loss. After the 1980s, deforestation rates steadily declined to 78 million ha in the 1990s and 52 million ha in the 2000s.[27] However, several reports now suggest that global deforestation has risen to anywhere between 100 million and 260 million ha per decade – that's more than ten times the size of the UK.[28] Some studies suggest we cut down around 15 billion trees each year.[29] If you could stack the felled trees on top of one another, they could reach the moon 392 times over.[30] It's difficult to know how many trees are planted each year, but many people report in the region of 1–5 billion.[31] Sadly, though, many of these trees, particularly in urban areas, will lack the care they need to thrive. One report has indicated that in June 2019, deforestation rates in the Brazilian Amazon increased by 88% compared with the same month in the previous year.[32] The recent warming climate has dried out parts of the forest, making it easier to combust. Therefore, forest fires have contributed to the increase in deforestation.

Environmentalists worldwide have accused former Brazilian president Jair Bolsonaro of allowing deforestation to accelerate. He has even been accused of 'crimes against humanity' at the International Criminal Court for his alleged role in contributing to the Amazon's destruction. It's the first case seeking to link deforestation to loss of life. However, there might be hope in Brazil as the new president, Luiz Inácio Lula da Silva has pledged to end deforestation by 2030. This is a bold claim, and 'big ag' lobbyists (large agricultural corporations) and Congress might be working against it, but current indications are promising. The government recently stated that deforestation in Brazil's Amazon fell by 33.6% in the first six months of Lula's term compared with the same period in 2022. He faces immense challenges, though, as the area of forest lost during his term is still more than three times the size of New York City.

In summary, we've lost billions of hectares of forest over the years, primarily owing to human activities. And we continue to lose vast areas. The rates are increasing in some parts of the world but may be decreasing in others.

Here come the nuances

It's imperative to shed light on the nuances of this narrative – particularly when it comes to asking how much new forest is actually growing. Recent data suggest that despite the ongoing devastation from deforestation, 130.9 million ha of land gained tree cover between 2000 and 2020.[33] That's an area larger than Peru.

The World Resources Institute summarised the findings of this study, suggesting that three countries, Russia, Canada and the United States, have more than half of the world's total tree cover gain: between 2000 and 2020 this was 68 million ha. But all three countries still lost more tree cover than they gained. National statistics can also be misleading since these countries are huge and varied. For instance, the state of New England was almost completely deforested in the nineteenth century and is now one of the most heavily forested regions in the USA. My colleague told me: "Everywhere you walk in the forest there, you find the ruins of nineteenth-century farms."

On a brighter note, 36 countries gained more trees during that 20-year period than they lost. Indeed, Ireland, Tajikistan, Bangladesh, Poland, Denmark and the Netherlands had some of the biggest increases in tree cover. It's thought that continental Europe has more tree cover now than in 2000, with a net increase of 6 million ha. Moreover, as Fred Pearce recently pointed out in a *Science* article, trees are cropping up all around the Arctic Circle as the climate warms: "Globally, recent research indicates forests are expanding along two-thirds of Earth's 12,000-kilometre-long northern treeline – the point where forests give way to tundra – while receding along just 1%."[34]

It's not just the polar regions that are gaining forest cover. At lower latitudes, the increase of CO_2 in the atmosphere enables plants to thrive. This is because it acts as a fertiliser that piles on the biomass; it also increases water efficiency in trees as they don't need to open their stomata so wide to acquire the CO_2 needed to photosynthesise. Yet the ongoing warming is working against tropical forests. We're therefore witnessing a profound shift in the patterns of forest cover across the globe, with gains in the polar regions but huge losses in the tropics. Whether this is good or bad depends on perspective, but it's likely to affect societies and biodiversity in unimaginable ways.

Even though the world has witnessed a notable increase in tree cover over the past twenty years, it's crucial to acknowledge that the loss of forested areas currently surpasses these gains, resulting in a net loss of over 100 million ha.[35] Moreover, planting new trees cannot compensate for the depletion of old-growth forests, which are essential for storing carbon and supporting biodiversity. Nonetheless, these newly acquired data offer an opportunity to analyse the locations and causes of the observed gains, thus presenting fresh possibilities to monitor and guide global initiatives focused on forest restoration.

Additionally, millions of people depend on harvesting trees and forest products as a means of livelihood and as a source of energy. However, as a recent scientific publication highlighted, specious assumptions about deforestation drive ineffective bans on these practices.[36] It was claimed that banning people from low-intensity harvesting imperils the climate and poor communities. The authors argue that many different types of low-intensity forest harvesting can allow a forest ecosystem to flourish and even avoid harmful carbon emissions. Importantly, these low-intensity practices are vital to the livelihoods of many communities globally. However, they contend that these practices are "routinely conflated with the main drivers of emissions from deforestation". Essentially, the authors suggest that people in high-income countries who impose bans on tree harvesting in the Global South often ignore the importance of tree harvesting for the livelihoods and cultural integrity of local peoples. These bans can drive tree harvesting 'underground', with a consequent loss of regulatory control and criminalisation of local people who rely on the forests.

Tree diseases

The loss of trees and forests owing to plant diseases (or phytological diseases) is another form of deforestation, but one that's not typically referred to as such. Plant diseases are part of the natural order of the world, but some globe-trotting pathogens are inadvertently transported and spread across the world by humans. Moreover, the pressure trees are put under from various pollutants (air, light, noise and heat) can weaken their immune systems, thereby increasing their susceptibility to diseases. Unfortunately, tree-killing epidemics are on the rise.

Part of my day job is to venture out into the countryside to locate trees that were previously plotted on a map and then climb them to check for elusive animals – to make sure nearby construction developments are not harming them. I used to enjoy finding native ash trees, which are often encountered along a woodland edge or in the middle of a field. Some were tall and graceful, over 30 m high and with sprawling branches that allowed for excellent traversing of their domed canopy. Deep in the woodland, others were like giant beanpoles as their stems reached up towards the light. Ash trees have around eight green oval leaflets positioned in opposite pairs along a midrib with a terminal leaflet. The technical term for this configuration of leaves is 'pinnately compound'.

Ash leaf with its leaflets in opposite pairs.

In the summer, their canopies would light up hedgerows on the horizon, swaying iridescent green in the breeze. However, a few decades ago, a fungus from Asia, colloquially named 'ash dieback', arrived in Europe. In the fungus's native Asia, it doesn't cause a huge amount of damage to its native hosts, the Manchurian and Chinese ash trees. But native ash in Europe did not evolve with the fungus, so they have limited natural defence against it. This is much like the SARS-CoV-2 virus that causes COVID-19, where a novel pathogen encountered a naïve host (humans).

Unfortunately, even the stretch of water that separates the UK from mainland Europe couldn't prevent the fungus from reaching British shores, and in 2012, it was first recorded in the south-eastern forests of England. The spores of the fungus can travel in the wind like floating microscopic armadas, so it's possible that it could arrive naturally, but we now know it was also imported on ash saplings. Since then, ash dieback has wiped out hundreds of thousands of trees across the UK, and it's thought we'll lose around 80% of our ash trees in the coming decades. This will have a devastating impact on our landscape and biodiversity.[37] When I venture into the countryside now, it's often easy to see the signs of this change. Those green canopies lighting up the hedgerows often appear black and lifeless. As Aldo Leopold said: "One of the penalties of an ecological education is that one lives alone in a world of wounds."[38]

Thousands of similar tree diseases are taking the world's ecosystems by storm. The spread of tree pathogens reduces biodiversity, disrupts ecosystem functioning and harms economic development. In Africa, the accelerated spread of pathogens is threatening the livelihoods of human communities. Around 100 pathogenic species are affecting trees in plantations and natural forests across North and Central Africa.[39] Broadleaved trees are affected the most, with only one in three pathogens attacking conifers. The rate of non-native pathogen

The author climbing an aspen, a tall temperate-forest tree, in the UK.

introduction is increasing, and it's becoming more difficult to predict the next invasion as climate change affects how species disperse by altering Earth's ecological processes.

Imagine waiting for your post to arrive. In a harmonious system, you generally know that your bills, letters and parcels will arrive within a certain timeframe and in a predictable condition (although there are always exceptions). There's the odd anomaly – much as in nature – but generally it's predictable. But now imagine that owing to extreme societal upheaval the system starts to collapse: mail workers go on strike, the roads that delivery drivers use begin to disintegrate, arsonists set the mail-sorting factories on fire, some people's letters are doused in anthrax and havoc is the new norm (a perhaps somewhat sensationalised analogy for climate change – or perhaps not). It then becomes difficult to know when or even if your mail is coming, and if it does what condition it will arrive in and whether it's safe.

The invasion of African ecosystems by pathogens is a growing emergency on a continental scale. And unfortunately, other continents across the world are in a similar position. However, action is under way to reduce the impacts of plant pathogens on our forests. This includes developing better policies to control imports and exports, improved knowledge-sharing and innovative microbiological tools to enhance pathogen resistance in trees.

Spillover

Evidence also suggests that deforestation can increase infectious disease outbreaks.[40] As humans diminish biodiversity by removing forests, we increase the risks of pandemics such as COVID-19.[41] As we encroach on wild habitats, we increase the likelihood of direct contact with potential disease vectors such as insects, bats and birds. This can sometimes enable a pathogen to jump the species barrier from a non-human animal into humans. Scientists call this a spillover event. One that likely occurred owing to deforestation involved the Nipah virus. In 1998, this novel paramyxovirus emerged in Malaysia, causing a fatal disease in pigs and humans. It's thought that the expansion of farming brought pigs into areas where fruit bats, the natural hosts of the virus, lived, and the virus jumped from bats to pigs to humans. Additionally, slash-and-burn deforestation resulted in a severe haze that blanketed Southeast Asia in the months preceding the virus outbreak.

Evidence suggests that this, along with a drought, reduced the availability of fruiting forest trees, which led to bats seeking their food in the local orchards and increasing the disease transmission risk.[42]

There's also a contested theory called the biodiversity dilution effect.[43] This theory posits that a large number of different species reduces the risk of infectious diseases because pathogens are 'diluted' among many animal hosts, which differ in their capacity to infect invertebrate vector species such as ticks and mosquitoes. If this is the case, ensuring our forest ecosystems are healthy and diverse (while preventing encroachment) could be an important disease prevention measure.

Forest fires

Again, this phenomenon doesn't necessarily fall into the category of purposeful felling, but forest fires threaten forest ecosystems globally, and their increasing frequency is linked to human-caused climate change and poor land management.

For several millennia, the Indigenous Australians burnt their landscapes in a process known as cultural burns.[44] They did this to replenish the earth, enhance biodiversity, encourage flowering plants and create a mosaic of ecologies with different microclimates. A central notion was the starting of a 'cool fire'. This means initiating a fire at night or early morning when plants secrete fewer flammable oils and when night-time dew helps to cool the fire down. It's thought that cool fires don't destroy seeds and root systems and can even help soils retain more water. Controlled fires can allow animals to escape and improve vegetation structure by reducing the density of aggressive species. By controlled, I mean that the Indigenous Peoples used to stay with the fire to manage it. Their acute understanding of wind patterns helped them to control the flow and direction of the fire, while they used knowledge acquired over millennia to assess the ecosystem, the fuel loads, animal breeding times, soil and grass types, and the ashes the fire left behind. There were different burning styles, which depended on the condition of the land and the local ecology. These cool burns also preserved the canopy of trees. This is vital because protecting the canopy allowed animals to return swiftly, prevented vast amounts of carbon release and preserved the tree's natural lifecycle. Smoke from cool burns could also trigger germination in the trees – particularly in water-permeable seeds – thereby kickstarting a process of regeneration.

There are also naturally occurring forest fires. Lightning strikes the earth around 8 million times daily (or a hundred times each second), and many of these strikes can cause wildfires. They can also be caused by accumulating leaves and twigs that produce sufficient heat to combust and ignite the surrounding vegetation. These fires are increasing owing to climate change. However, according to several reports, most wildfires in the USA, Australia, Russia and the Amazon are caused by humans who, for example, drop cigarettes, use fireworks, burn debris, light barbecues and so on.[45]

Climate change only exacerbates these human-induced blazes. I remember as a child standing in the garden with my father burning some debris in the incinerator. One of the items was a large piece of old fabric. As we put it in the incinerator, a gust of wind blew it out of our hands and onto a row of cypress trees. The trees set alight immediately, and the whole row was scorched within minutes. In the right conditions fire moves fast, and if trees are dry or contain high levels of flammable oils, the wrathful flames wait for nobody.

Hundreds of millions of hectares of forests are now lost to wildfires each year. In Australia, 24 to 34 million ha of forest and grassland ecosystems burned during the devastating 'Black Summer' season of 2019–20.[46] But we can and must do something about these increasing forest fires across the globe. Re-evaluating forest restoration strategies, understanding soil ecology better, working with Indigenous Peoples and promoting their leadership and sovereignty, and changing our intensive land management practices to reduce droughts are some of the things we need to do.

If we work with nature rather than asserting dominance, we can pave the way to a more resilient and hopeful future.

Petroleum – I Once Was a Tree

Petroleum is a fossil fuel formed from the remains of ancient trees, plants and other organisms that lived millions of years ago. Over time, these organic materials are transformed into crude oil and natural gas deposits found deep underground. It's the carbon in trees and other plants that eventually contribute to petroleum formation. When plants die and decompose under specific geological conditions, their organic material can undergo a process called diagenesis, where it's buried and subjected to heat and pressure over millions of years. This results in hydrocarbon-rich substances.

Trees' contribution to the carbon cycle and their ability to sequester CO_2 through photosynthesis play important roles in mitigating climate change. Forests help remove CO_2 from the atmosphere, storing carbon in their biomass and soil. This is critical in addressing greenhouse gas emissions and maintaining a balanced global carbon cycle. So, while ancient trees contribute to petroleum, we must keep these fossil fuels in the ground and allow living trees to flourish. Fossil fuels have driven the technological world we live in today. Trees have shaped my ability to write this book on my laptop and research the content of these pages using the internet. They've allowed me to communicate with my publisher and with other people on social media. They've shaped the very structures and function of the modern world.

CHAPTER 3

The Social Importance of Trees

"And there I marvelled at the forest's trees; my soul rooted to its deep verdant sea. I could no more leave it than it could walk from me."
—Angie Weiland-Crosby

I'm walking through a broadleaved woodland to record my thoughts for this chapter. I hold a dictaphone in one hand and gently brush the moss-covered trunks with the other. I'm comforted by the musky scents, diverse fractal patterns and the menagerie of green entities surrounding me. My microbial ecology research has taught me that I am far from alone on this journey. I exchange microbial residents and other biological molecules with those of the trees, the soil and the animals whizzing through the air and scurrying in the leaf litter. I leave a part of me in this wholesome space and gain something biologically new in return – a cloak of forest microbes and organic compounds. But it's not just a biological affair; it's also educational. I invariably learn something each time I walk through a woodland, whether it's a better understanding of forest symbioses or discovering novel insights into the feeding and breeding behaviours of the avian forest dwellers. I know these tacit lessons have downstream influences on my life in weird and wonderful ways – for instance, novel research ideas, drawing parallels with human health, or strengthening my respect and awe for the complexity and beauty of other organisms. Indeed, the lessons feed my desire and ability to protect nature in return.

It's also a psychological affair. All my worries are left behind, gradually dissipating as I embark on my forest journey. The ensemble of chemicals that tinker with my brain in anticipation of these organic connections certainly plays a role. My mood elevates and my attention increases.

This is a personal account of why trees are socially important. It's a story of connection and, in a way, of companionship. But there's something more fundamentally important from a social perspective. Trees play a pivotal, life-sustaining role in all societies on Earth. It's virtually impossible to overstate their social importance. They support our health, hold a central place in many spiritual beliefs and are vital for human livelihoods.

Shinrin yoku

In the 1970s and 1980s, Japan's economy was booming thanks to rapid technological advances and clever policies implemented after the Second World War. For instance, the country had captured over 20% of the world's automobile market,[1] and was a world leader in miniature technology, such as microchips and circuit boards. This remarkable technological prowess led to an era affectionately dubbed 'Japan Inc.'. Japanese people were working hard, earning healthy sums of money, and their cities were expanding and becoming more populated. However, this technological and economic boom came with a price. Japanese workers were feeling the burn. Doctors started to notice a rise in symptoms such as depression, anxiety, headaches, body pains and attention deficit. It can be immensely challenging to truly relax and find solace in cities. Sensory overload is caused by heavy traffic, bright lights and noises, and spending long hours in the office adds to the stress. Over time, the high demand for real estate as urbanisation increased left many towns and cities depleted in nature, with few trees and green spaces to bestow even the illusion of a natural setting.

Japanese physicians and researchers began to investigate forest therapy in response to the increasing stress levels and the spike in chronic, non-infectious diseases that were ravaging urban dwellers across the country. *Shinrin yoku* was developed, a concept based on ancient Shinto and Buddhist practices.[3] This means to bathe in the forest and let nature into your body and mind through all five senses: seeing, hearing, touching, smelling and tasting. The main aim of shinrin yoku was to halt the ever-increasing maladies of Japanese workers and restore their health. Presumably, the government also envisioned benefits in productivity terms. Indeed, the practice of shinrin yoku also aimed to revitalise the economy of rural areas, which were suffering from the mass migration of people into the cities.

In the late 1980s, Yoshifumi Miyazaki was commissioned by the Japanese Forestry and Forest Products Research Institute to research the restorative properties of nature engagement – why do we feel calm and nourished when we encounter nature? The first government-backed shinrin yoku experiments were carried out on Yakushima, an island off the tip of Japan's southern mainland. Here, Yoshifumi researched the impacts of cedar forests on stress hormones in the human body. He found that people felt less stressed when engaging with the forest environment and that this could be linked to the regulation of stress hormones.

Cue Qing Li at Nippon Medical School in Tokyo and author of *Forest Bathing: How Trees Can Help You Find Health and Happiness.*[4] Li has published several scientific studies on the effects of shinrin yoku on various health indicators, including blood pressure, stress and immunity. Li estimates we spend an astonishing amount of our lives – some 93% – indoors. This, he suggests, is having a tremendous effect on our health and wellbeing. After all, we evolved to be surrounded by diverse life forms, not cooped up in lifeless cubes. Sure, we've always had shelters of some variety, but only recently, in evolutionary terms, have we been shutting biodiversity out of our lives as we stay largely static in our houses and workplaces. Yet we need exposure to biodiversity to ensure our 'walking ecosystems' flourish.

Li states that around 62 forest therapy businesses in Japan provide a health service via forest therapy guides.[4] Japan even has a word, 'karoshi', which means 'death by overwork'. Li says it makes sense for people to turn to the forests to help them thrive. He suggests that shinrin yoku can reduce blood sugar levels and stress hormones, lift depression and improve sleep quality and concentration. The practice involves drawing in the forest environment through all senses – walking through a forest and taking the time to notice what you see, such as the fractal patterns of the trees. It involves mindfully breathing in the forest air and feeling the rich textures of the trees' undulating bark. It involves noticing the rich earthy aromas that stimulate the olfactory system (the part of the body responsible for processing smells) and listening to the wind, the shimmering leaves and the birdsong. It's a truly multisensory experience that requires you to surrender to and embrace your environment. The various stimuli in

The lush bamboo forests in Japan are a popular option for forest bathing.

the forest environment can trigger activity in different parts of the brain, leading to the secretion of neurotransmitters and hormones that eventually slow your heart rate and make you feel calm. There's even evidence that organic compounds called phytoncides emitted by the trees and floating around in the air can interact with the immune and central nervous systems in beneficial ways.[5] These essential oils have been linked to reduced blood pressure and enhanced immune function. One study has also suggested that phytoncides can interact with neurotransmitters in the central nervous system and improve sleep, in mice at least.[6]

While Japan is widely credited with popularising shinrin yoku, the concept is far from novel. Many cultures have long recognised the restorative importance of mindfully engaging with forest habitats. Yet modern science is now catching up, and more attention is being given to understanding the mechanisms involved.

Other health and wellbeing benefits

My own research has shown that trees may contribute to human health and wellbeing through another biological pathway.[7] Trees are giant vectors, emitting millions of microbes into the atmosphere every minute. These invisible organisms bombard humans on a daily basis. Studies have shown that up to a million microbial cells can be found in a single cubic metre of air, and people can inhale an astounding 100 million bacteria each day.[8] Crucially, it's thought that exposure to a diverse assemblage of microbes from a young age is important for our long-term health. For instance, microbes from the environment train and 'educate' our immune systems.[9] Exposure to a diverse environmental microbiome helps to build a repertoire of immune cells (much like a tiny army) that can recognise and fight off pathogens. However, a fear of all microbes is spreading, and we're shutting biodiversity out of our lives.[10] People in deprived neighbourhoods may be at higher risk of being exposed to disease-causing organisms.[11]

Certain microbes in the environment are 'old friends' because we've co-evolved with them over thousands of years.[12] They have shaped our immune systems and play fundamental roles in regulating our innate immunity – a stalwart defender of the body. This innate immune system can be compared to a security guard stationed at the entrance of a building. Just as a security guard screens visitors at the door to ensure that they are authorised to enter and pose no immediate threat, the innate immune system acts as the first line of defence against pathogens that enter the body. It's always present and ready to respond to any potential threat. It's not specific to any particular pathogen and doesn't require prior exposure to the pathogen to mount a response. This makes it a crucial component of the immune system, providing immediate protection against a wide range of pathogens. So it's likely that we need to be exposed to a swathe of different microbes from environments such as forests to keep our ecosystems healthy.

Microbes also possess important genes that play roles in sustaining our body's ecosystem. A recent study I was involved in suggested that many life forms floating in the air originate in the trees and the soil beneath our feet.[13] This makes a lot of sense. Soil is arguably the most biodiverse habitat on Earth, and a single gram of it can contain more microbes than humans on the planet.[14] Trees are deeply connected to the soil physically and biologically through chemical exchanges with soil-dwelling microbes.

Microbes are extremely light, so they become airborne easily and are carried far and wide by the wind. They can be lifted from the trees and the soil in tiny air bubbles that form in raindrops and clump together on dust particles that fall from the atmosphere. Our study found that the air in forested urban parks in South Australia contained more bacterial species but fewer potential human pathogens than nearby sports fields or less wooded environments. Trees appear to filter the microbial communities in airspace, reducing the risk of exposure to disease-causing microbes. Because trees also seem to increase microbial diversity (the number of different species) in the air, allowing more trees to grow in urban areas could provide an important health benefit by enhancing our immune systems.

Despite not being able to see the microbes and the other members of the invisible world around us, they're fundamental to the proper functioning of ecosystems.[15] Therefore, spending time in and around trees is immensely important from a social perspective. Human and ecosystem health are deeply interconnected with our happiness and ability to function as part of a flourishing society.

Spending time in forests is also the subject of a concept called green prescriptions, also known as nature-based health interventions.[16] Much like normal medical prescriptions, green prescriptions are given to people with health-related conditions. These could be cardiovascular diseases or mental health issues, or even social isolation – a leading risk factor for mortality, one that has been likened to the habit of regular smoking. A doctor essentially prescribes people time to engage with the natural world. This may seem strange, but owing to increasing social barriers and the fact that access to quality 'greenspaces', including forests, is being reduced, people often need a structured programme that guides them to reconnect with the natural world. Many organisations now run forest-based green prescription programmes. These can be highly social in themselves: you go out to the woods with a group of likeminded people and make friends. However, the activities should be tailored to suit an individual's needs and interests to be successful… Let the forests be thy medicine.[17]

Several studies have shown that increasing tree cover in a neighbourhood can lower the risk of depression and enhance sleep quality.[18] Trees are also important in cooling the urban environment, thereby increasing human comfort.[19] When the sun shines on a tree, it absorbs energy and uses it to power photosynthesis. But not all the energy is absorbed; some is reflected into the atmosphere. Trees also release water vapour through tiny pores on their leaves, which helps to cool the surrounding air as the water evaporates.

The 'urban heat island effect' occurs when natural land cover is replaced with roads, buildings and other surfaces that absorb and retain heat. It's a form of localised climate change that can increase air pollution levels, heat-related illness and mortality. Heat is estimated to cause over 100,000 patient days in hospital annually in the UK alone.[20] Urban trees provide a vital cooling effect, and are therefore another benefit to society.

My colleagues and I recently published a paper that explored the differences in tree cover in the centres of British cities and large towns.[21] A focus on city centres was important for several reasons. One was that owing to the urban heat island effect, temperature intensity is typically highest in the core of a city. This means it's vital to ensure plenty of healthy trees are growing in city centres: in addition to their intrinsic value, they keep the city cool.

We discovered a north–south divide in Great Britain, with tree cover (and other green features) significantly greater in England's southern cities than in northern ones. Seven of the ten *least* green cities were in the north and seven of the ten greenest cities were in the south. The reasons for this north-side divide are complex and to an extent depend on decisions made centuries ago – some urban centres invested in parks and tree-lined avenues more than others in the past. Industry and war efforts contributed to urban sprawl and reduced natural features in some cities, particularly in northern England and Scotland. In the nineteenth century, city planners often incorporated street trees, particularly in affluent areas. This decision was influenced by an admiration of continental European boulevards and a recognition of the wellbeing benefits of garden cities and spa towns. Related is the 'luxury effect', whereby affluent neighbourhoods record higher biodiversity in cities worldwide, something that's often dictated by structural classism and racism. Therefore, differences in historical development have left us with leafy urban centres such as Exeter's (which came in at number 1 in our study) and others with far fewer trees and greenness (Glasgow's city centre was bottom of the table). The question remains why over time such imbalances have not been addressed. But given the social importance of trees, we need to do so as a priority.

Sacred tree groves

Sacred groves are a legacy of prehistoric nature conservation traditions. They may seem like patches of forest with no social connection, but the rural communities that protect and revere them as sacred are spiritually connected to them. The beliefs of these communities ensure that these groves and even the creatures that call them home are protected from harm. This protection means these areas are often the last refuge for rare species.

Once upon a time, sacred groves were widespread across the globe. They were reported in Sanskrit and Ancient Greek scriptures but were largely wiped out by Europeans as colonialism was driven by the resources it brought. However, examples are still found in parts of Asia, Africa and the Middle East. As Western concepts of wild places free of human activity were imposed worldwide, conservation organisations created 'protected areas' and Indigenous communities were evicted from them. This neglected the fact that in some cases these peoples have co-evolved with and nurtured sacred groves for millennia, protecting them from harm – the very thing that top-down conservation is supposed to promote.

Small sacred groves are scattered across India, often protected by rural villages – both for religious reasons and for the rich bounty of provisions forests confer. Early British colonists described India as an 'ocean of trees', and forests were rapidly seized from the local villagers by the English East India Company; they became state property. Many sacred groves were removed to expand agricultural fields, but the locals continue to fight to protect the tens of thousands of these tiny forest fragments that are thought to remain. The groves are sites of pilgrimages for several Indian-origin religions, including Hinduism, Buddhism, Jainism and Sikhism. Local art and folk traditions are connected to the deities of sacred groves. Indeed, these forests are a vital facet of many societies.

In 2010, I decided to visit the coastal rainforests of East Africa. I travelled to Ethiopia, Tanzania and finally Kenya. I was away for around three months and worked on various rainforest research and community sustainability projects. My abode for most of the trip was a clay house in the tiny fishing village of Shimoni, a village located in south-eastern Kenya and surrounded by biodiverse coastal rainforests. The area is classified as a 'biodiversity hotspot' owing to its many endemic species (those restricted to a geographical area that do not occur naturally in any other part of the world). The communities of Shimoni and other local villages depend on their forest for firewood, materials to build their homes and canoes, and traps for fishing.

The coastal rainforest was equally immense and inspiring. I fondly recall lying in a hammock, my ears acutely attuned to the sound of the forest's insects and the orchestra of stridulation (rubbing their body parts together to communicate) that dominated the evening soundscape. It was a rich cacophony of nature's vibrations in an endless maze. I was coaxed out of the confines of my sensual comfort zone and thrown into a profound theatre of delight. The sound was deafening yet strangely relaxing at the same time. My nose was filled with diverse odours from the soil, the air and the trees. As I swung in my hammock, my eyes focused on the largest row of ants I'd ever seen, many thousands of them scaling the trees. I started to notice the simple fact that all the stimuli tickling my various senses were supersized compared with those in my usual environment. It was a special place.

During my trip, I spent considerable time with the Shimoni villagers recording the biodiversity of the forests and observing the behaviour of the Angolan black and white colobus monkey. I was entrusted with a large, sharpened machete to chop down the understorey vegetation as we made our way into the realm of the deep forest. My guides educated me on the awe-inspiring diversity that surrounded us; they taught me how to call the monkeys and track elephant shrews. Once, a large, shaggy vine hung down from

The black and white colobus monkey, like those in Shimoni's coastal 'coral rag' forest.

the canopy. To me, it was just another vine, but to my Shimoni village friend, it was far more than that. He bent the vine above our heads and made an incision into the centre. After a couple of seconds, water started pouring out as he held his mouth open to drink. That the vine was a source of fresh water was invaluable ecological knowledge in the deep and humid forest where dehydration can strike at any moment. Despite my new-found knowledge of the water-giving vine, on subsequent days I still carried many litres of water, not trusting myself to identify the vine again correctly.

Clearly, the local peoples were deeply interconnected with their forest and held ancient wisdom crucial to their livelihoods. One day, I was told that the Shimoni forest also contains 'kaya' forests with sacred shrine trees where ancestors were buried. I was asked if I'd like to visit one.

Machetes in hand, we hacked our way through the vines. Among the wonderful creatures we encountered on the way – such as a pit viper and silvery-cheeked hornbills – were ants with a penchant for human flesh, and clothes were no deterrent. My guide had spotted the army of them from several metres away as they crossed a lightly defined path in the forest. We were told to either find our way around or run like the wind and jump over them. I chose the former option and found myself on the other side unscathed. My friend, however, decided to run and jump but tripped and fell. The ants covered him in seconds. According to the guide, the best thing to do in this situation was to move as far away from the ants as possible, strip off all your clothes and try to shake the ants off your body. Who were we to question? My friend took heed of the advice. I'll leave the rest of the scene to your imagination.

Once in a while, we encountered a strange type of rock beneath our feet. It was sharp, gnarly and covered in holes. It was coral rag – a rubbly limestone composed of fossilised coral reef. Most of the coastal forest had established itself on these mineralised remains from millions of years ago. It was certainly a strange feeling to be walking through a rainforest with fossilised coral poking out of the ground.

We eventually arrived at the tree shrine. It was a large deciduous tree with giant leg-like buttress roots at the base. In the spaces between the roots were many items – mostly glass bottles containing a liquid (possibly rosewater) and other gifts. I was told that village elders continued to respect centuries-old rituals and still leave honey and rosewater offerings to this day. It was an interesting experience, and I can understand why these immense trees are important spiritual places.

On the way back, my Shimoni friend taught us about the baobab tree and its giant fruits. The baobab symbolises resilience, having adapted to harsh and changing environments over thousands of years. Its massive trunk can store large amounts of water, allowing it to survive in arid regions where other trees cannot. Its thick bark protects it from wildfires and other forms of damage, and its long taproot helps it to find water deep underground. As we returned to camp, I noticed many of the baobab's giant fruits were spread across the forest floor. Some were the size and shape of a rugby ball: I certainly wouldn't want to stand under the tree as the fruit dropped. The famous 'upside-down tree' fruits are tangy, dry, sweet and full of fibre and vitamins. This nutritious fruit is another reason the forests are important to the local community.

The villagers continue to work hard to protect the coastal forest of Shimoni. They face tremendous pressures but hope ecotourism can help conserve the forest and its biodiversity. It's a beautiful place with mangroves, beaches and great snorkelling sites. The forest and the local people who depend on it are equally wondrous.

Ceremonial and cemetery trees

In Western Europe, one tree is associated with cemeteries more than any other: the yew. This long-lived tree has been firmly regarded as a symbolic and mysterious being for millennia. In the UK, the druids (members of the high-ranking class in ancient Celtic cultures) upheld the yew tree as a sacred entity integral to death rituals, so much so that Catholic authorities in the sixteenth century petitioned to ban yew trees in England.[22] The yew's mystique is firmly ingrained in several cultures. For instance, it was sacred to the ancient Greek goddess Hecate. In Greek mythology, yew trees were seen as important purifiers of the dead before they entered the underworld.

But why does the yew tree symbolise death and sometimes resurrection? It may be because of its cyclical and regenerative nature: yew branches can take root and form new trees. These trees have been tied to church environments for hundreds of years. Planting them also deterred livestock from trampling on the graves; indeed, the yew tree is highly poisonous to cattle and humans alike, with the bark, seeds and needles all being toxic – only the red fleshy berry is not. The seventeenth-century botanist Robert Turner mentioned the belief that yew trees were poisonous because they 'run and suck nourishment' from the dead.[23] In reality, the toxic components of yew trees are taxine alkaloids. These compounds are absorbed quickly from the intestine, and in high quantities they can cause cardiac failure, cardiac arrest or respiratory failure. The yew tree was also important to medieval militaries, but not for its poison. Until the advent of firearms, the longbow was one of the most formidable weapons of medieval times, and it was made of yew, widely regarded to be the best bow wood in the world.[24]

Some churchyard yew trees are huge and are thought to be thousands of years old. The one at St Cynog's in Powys, Wales, is reportedly 5,000 years old.[25] The Crowhurst yew in Surrey has a famous split trunk, and in the nineteenth century a table and chair were added to the tree's interior. The yew tree at St Anne Chapel in La Haye-de-Routot in northern France is over 1,000 years old and has a room inside. I was lucky to visit the yew tree in the churchyard of Hayling Island, Hampshire. It has a notice board next to it describing how the tree was alive in the days of Jesus Christ. This is very hard to believe, but it's certainly many hundreds of years old. When I looked up into the dense branches, it was so large that it felt as if I was peering through the canopy of an entire forest.

On another visit to a churchyard near Gosport, Hampshire, my wife Kate once saw what she thought was a holly tree in the distance. Upon closer inspection, it was indeed a holly, but it had grown inside the shell of an ancient yew. Even after the yew had died, it provided a nurturing shelter for a wind-blown holly seed to grow into a mature adult tree.

This phenomenon is a strange example of epiphytic growth or epiphytism, in which the holly uses the yew as a support structure. Epiphytic growth occurs when a plant attaches itself to a host tree and begins to grow, obtaining nutrients and water from the surrounding environment. Over time, the holly may grow and spread throughout the yew tree, utilising the decaying matter of the dead tree for additional nutrients. To me, this was the ultimate symbol of strength and fortitude – a presence, forever aglow, from one life to another.

In Australia, the Indigenous Peoples mark important ceremonial sites by carving intricate, ornate patterns on tree trunks.[26] They did this for thousands of years before the arrival of the European colonists. The carvings are exquisite, comprising swirls and zigzags that represent symbolic motifs. Although this was once a widespread practice, it's far less common now: at the time of colonisation, Indigenous Peoples were moved into reserves and forced to accept European-style tombstones instead. There were once thousands of these beautifully carved trees dotted across the landscape. Most were destroyed by land clearance, but a few were removed and stored in museums: a number have been traced to museums in Geneva, Oxford and Melbourne, and they are also housed in private collections. Some communities are

The churchyard holly tree growing inside an old yew.

trying to bring the trees back to their homeland. In 2010, Museum Victoria returned a carved tree to the Baradine Peoples in north-west New South Wales, Australia. It's my hope that the other carved trees are returned to the Indigenous communities across Australia from which they came. I recently watched a film called *Gaaguuwiya Dhawunga* (Bring Back Home).[27] It's openly available on YouTube, and I encourage you to watch it. The movie documents how thousands of carved trees of the Banarway Bora, which played a crucial role in

Indigenous knowledge systems (having 'a spirit that is past, present and future'), were cut down and sold to museums worldwide, and the filmmakers rightly call for their repatriation.

These trees have profound spiritual meaning for the Indigenous communities, and they're calling upon public and private collectors to return them as part of an ongoing healing process. Much like the yew tree mentioned earlier, these trees still have the power to nurture other organisms (in the former case, it was a holly tree, and in this case, it's human communities) even after the trees have passed away.

The politics of urban trees

I consider myself lucky to have gone to the University of Sheffield, Yorkshire, to do my PhD. Sheffield is now a city of around 600,000 people. It has a long industrial history and was the centre of steelmaking in England, from the fourteenth century, for several centuries. Sheffield is famously built around five rivers, and the wider Sheffield region is often considered one of the greenest urban areas in Europe. It has 80 public parks, 650 green spaces and over 2 million trees. This means it has more trees per person than any other UK city region, with trees outnumbering people by nearly four to one. For comparison, there's one tree per person in London. The trees are scattered across Sheffield's streets and in the 170 woodlands that cover 28.27 km^2 of the land.

If you hopped onto a train and travelled straight into Sheffield's city centre, you'd probably scratch your head after reading the previous few sentences. Strangely, there aren't many trees there. As mentioned earlier, I recently published a study with my wife (and data scientist) Kate, Paul Brindley and Suzanne Mavoa on how green British city centres were. The study showed that Sheffield's is one of the least green (of cities with over 100,000 people) in the entire UK.[28] This was equally surprising and concerning to us, and we hope our publication motivates the local authorities to improve tree coverage, which I know they're working on. As previously discussed, trees and other green infrastructure provide a plethora of health and wellbeing benefits for us, also providing supportive environments for recreation, conviviality and creativity.

Many people from diverse backgrounds spend considerable time in city centres, for employment, shopping and recreation. These are places where populations from otherwise socioeconomically disparate areas merge and mingle. I'm sure that, despite Sheffield's wider region including wonderfully enchanting woodlands, the people who live there would like to see more trees in the city centre. And speaking of the importance of urban trees, Sheffield was in the news recently when communities gathered to protest the felling of veteran street trees.

The Sheffield tree-felling protests

In 2012, a £2.2 billion 'Streets Ahead' contract was signed by Sheffield City Council, Amey plc and the Department for Transport. This was supposed to be a 25-year contract for road and pavement upgrades across Sheffield. The city council planned to remove a staggering 17,500 trees as part of the contract, considering the trees to be either dangerous or 'in the way'.

In 2014, the council planned to fell a 450-year-old oak in the Stocksbridge area of Sheffield because it was 'diseased'. This caused a public uproar, and despite local protests and an expert saying the tree was 'uncompromised', the grand old oak was felled anyway. Later that year, residents in another part of Sheffield campaigned to prevent the felling of 189 mature trees to make way for a bus lane. Their campaign was successful. Tensions began to bubble up across Sheffield because the street trees meant so much to local people. In 2015, several local tree campaign groups were formed across the city, and the city council launched an independent tree panel to advise on whether trees could be saved. Many of their concerns were ignored, and by the end of 2015, over 3,000 trees had been felled.

In early 2016, the High Court issued an interim injunction that ordered the council to pause felling for three months. However, it recommenced after this period, and the local police force was involved for the first time. In November 2016, during a peaceful protest, two people were arrested under Section 241 of the Trade Union and Labour Relations Act (1992), designed to prevent pickets from disrupting workplaces. The charges were hotly disputed by the tree campaign groups. Later in November, eight trees were felled on a street in Sheffield at 4 am! Despite the council clearly trying to work while protesters were in the land of nod, two people were still arrested. This sparked nearly 10,000 new Sheffield Tree Action Group members, who campaigned to save Sheffield's trees.

Throughout 2017, trees continued to be felled and local people continued to be arrested for protesting. In March 2018, there was a year-long pause, and in February 2019, seven campaigners arrested in 2016 and 2017 were awarded wrongful arrest payouts. A year-long investigation found that the council had deliberately misled the local community over the tree felling; politicians have referred to it as 'environmental vandalism' and a 'national scandal'.[29] In 2021, a pledge was signed by the council and the local conservation groups to ensure a shared vision for the city's trees is considered and acted upon.

The Sheffield protests epitomise the social importance of trees. A fellow Sheffield PhD graduate, Camilla Allen, recently co-edited a fantastic book called *The Politics of Street Trees*, in which the Sheffield tree protests are eloquently described, among many other political musings.[30]

The social importance of mangroves

Mangrove forests are characterised by groups of trees that live along the coastlines in tropical and subtropical latitudes. There are approximately 70 species of mangrove trees, and all have evolved adaptations to allow them to thrive in salty oxygen-deficient soil. They stand moderately tall, with their tangled roots stretching deep beneath the brine of the intertidal zone.

When Italian merchant Amerigo Vespucci (from whose name 'America' is derived) visited the South American coastline in the late fifteenth century, he named present-day Venezuela, which translates to 'little Venice'. The reason for this is that the dwellings that sat on stilts above the saltwater in the mangroves reminded him of the Italian town's canals and dwellings. The mangroves were vital for the livelihoods of the Indigenous Peoples of the time and are still today. They form barriers against severe storms and tsunamis, thereby saving lives and protecting dwellings.

Mangroves provide local communities with ample food, such as fish, crustaceans and fruit, along with medicines and wood. They prevent erosion of the shoreline by trapping sediments and building substrates. Indeed, mangroves are master trappers, so much so that they collect myriad plastic items that collect along the coastline. Taking an optimistic perspective, this could help people to recycle plastic waste. Indeed, during my time in Africa, the locals made the best of this situation by collecting the flip flops and sandals that washed up in the mangrove forests and upcycling them. There were thousands of recycled flip-flops for sale on the streets of Kenya. Recycling trapped flip flops doesn't provide a massive local income, but researchers estimate mangrove forests globally provide human communities with many billions of dollars' worth of so-called ecosystems services.

Unfortunately, recent mangrove deforestation has created hazardous conditions for coastal communities. Mangroves are chopped down to make way for urban development and shrimp farms. Seventy years ago, many Indo-Pacific coastal dwellers had 8 km of mangrove forest between the ocean and themselves, but today many villages are situated at the coastal edge, making them highly susceptible to typhoons and cyclones. One study found that mangroves can reduce storm-related mortality by around two-thirds.[31] This highlights the tremendous social importance of mangrove forests and the need to conserve and restore them globally. They are a vital link between land and sea, allowing a fragile balance that's essential for life's harmony.

Cork – I Once Was a Tree

Did you know that cork, that versatile material, comes from the bark of cork oak trees? It's a remarkable process that combines sustainability with practicality. Cork oaks, scientifically known as *Quercus suber*, are primarily found in the Mediterranean region, where they have thrived for centuries. The unique feature of these trees is their thick, spongy bark, which can be harvested without harming the tree itself. Every 9 to 12 years, skilled harvesters carefully remove the outer layer of bark by hand, allowing the tree to regenerate new bark over time. The harvested cork is then processed, cleaned, and transformed into a wide range of products.

One of the most well-known uses of cork is as a wine bottle stopper, providing an airtight seal to preserve the quality and flavour of the wine. Additionally, cork's natural properties, such as its elasticity and thermal insulation, make it a popular choice for flooring materials: it's comfortable to walk on and helps to regulate temperature.

The next time you encounter a wine bottle stopper, step on a cork floor or stick a pin in a cork board, think about the remarkable journey of cork – from the bark of oak trees to its functional applications in our everyday lives.

CHAPTER 4

The Ecological Importance of Trees

"One thing I've learned in the woods is that there is no such thing as random. Everything is steeped in meaning, coloured by relationships, one thing with another."
—Robin Wall Kimmerer

I'm walking through a patch of ancient deciduous woodland carpeted with bryophytes and minuscule wildflowers that may be missed even by the trained eye. It's a slightly misty August afternoon in 2022, the peak of the British summertime. When I arrived in the woodland, my auditory cortex was treated to an avian symphony, the perfect inspiration for this chapter's first topic – wildlife and ecology.

Wildlife supporters

The first thing my ears detected was the complex song of the tiny wren. In this moment, the wren was the concertmaster – it's always surprisingly loud for such a tiny creature. It cuts through other sounds like a hot knife through butter – its commanding vibrations bound towards your cochlea with unwavering determination.

I notice the intermittent call of magpies, which always reminds me of a slightly sinister cackling laughter, "hee hee ha ha", perhaps reinforced by years of bad press. They have a reputation for being egg-stealing vermin, when in fact their real beauty lies not only in their multi-toned appearance but also their immense success as a species. Magpies have evolved a remarkable level of intelligence and even the ability to recognise themselves. This latter is a rare cognitive trait found in only a handful of animal species, including great apes, dolphins and elephants. It contributes to their complex social behaviours, allowing them to flourish in different environments.

Many other bird species that live in our urban areas reside in the woodlands. Indeed, hundreds of bird species use forests as their primary habitat in the UK. And approximately 5,000 bird species call tropical rainforests their home.[1] The rich birdlife found in temperate and tropical forests emphasises the importance of these habitats for avian diversity.

The relationship between birds and forests is complex, encompassing mutualistic interactions such as seed dispersal and pollination. In many ways, forests don't just support wildlife; wildlife is part of the forest. Many forest plants have evolved to rely on specific bird species for seed dispersal. Fleshy fruits serve as enticing rewards for birds, which consume the fruits and transport the seeds to new locations. As the seeds pass through the bird's digestive tract, they are often deposited intact, accompanied by a small amount of fertiliser in the form of guano. This process allows for the colonisation of new areas and contributes to the plant's dispersal and genetic diversity.

I come across a fallen tree clothed in shaggy moss. Behind the logs, I find seemingly introverted fungal fruiting bodies – 'frosty bonnet' mushrooms. It's almost as if they're huddling together and hiding behind the mossy log. I spend a moment thinking about how these fruiting bodies are just the tip of the iceberg, albeit an essential tip. Attached to these bodies and below the ground lie dense mycelial networks. They weave their way through the substrate, each strand connecting to the others, the soil or trees, helping organic matter to decompose, potentially assisting trees to communicate and providing a food source for invertebrates. It has been said that each time you place your foot on the ground in a woodland, you could be standing directly above 300 miles of mycelium.[2] Indeed, mycologist Paul Stamets has said that just 16 cm^3 of myceliated soil can contain 13 km of hyphae cells – the branching filaments of the mycelia.[3]

Moss is another world within the forest ecosystem, remaining vivid green even as the leaves fade and fall in the autumn. As Robin Wall Kimmerer explains: "One gram of moss from the forest floor, a piece about the size of a muffin, would harbour 150,000 protozoa, 132,000 tardigrades, 3,000 springtails, 800 rotifers, 500 nematodes,

400 mites, and 200 fly larvae. These numbers tell us something about the astounding quantity of life in a handful of moss."[4] Forests are home to trillions of these invisible organisms, which I consider to be wildlife – just on a microscopic level. They reside in the soil, in the air, in the tree trunks, in the leaves and in the animals. Together they make up a vast network of forest microbiomes, each playing an integral role in the forest ecosystem. Invisible biodiversity is inordinately important to the forests' visible biodiversity.

Within the realm of a single tree, hidden wonders abound, surpassing what the eye can initially perceive. Beyond the leaves and branches, a world of insects and arachnids buzzes with life. Ants march, aphids suck sap, leaving a tell-tale trail, spiders spin their web of wonder and hoverflies pit-stop in the shade. Once you calibrate your eyes and ears, the tree becomes a vivid place to be, with its kaleidoscopic scattering of colours and theatrical sounds from the resident crawlies.

I recently finished collecting data for a study on tree insect communities. The fieldwork consisted of getting into my harness and rope-climbing up mature sycamore trees to record the various invertebrates living at different heights in the trees, from the lower trunk to the canopy tips. Spending time in the tree canopies is like going to the best school in the world. You can learn so much simply by observing nature. I observed variations in the abundance (i.e., number of individuals), richness (i.e., number of different species) and composition of invertebrate species based on their location within the sycamore trees. This investigation revealed that trees play host to a diverse array of insects and arachnids, each occupying distinct niches at varying heights within the tree. Certain regions of the trees were found to offer more favourable resources for specific invertebrate species than others.

Recall the beginning of the book, where I introduced the concept of the Umwelt (the perceptual worlds of organisms). Well, much as organisms have unique sensory niches, they also have unique food and shelter niches that determine where they live in a tree. I found that the invertebrates living on the underside of leaves were numerous, especially different kinds of aphids, ladybird larvae, and tiny (1–2 mm) pure-white spiders. If disturbed, the spiders can shoot their silk out and rapidly float on the wind to far-away branches. Other pure white insects were plentiful, too, much as in caves, where the animals lose their coloured pigments owing to a lack of light.

This loss of pigmentation is like the reverse effect of a fading painting in the light. A painting fades if left in the sun for too long, losing its vibrant colours, and animals exposed to low light levels over evolutionary timescales lose their vibrant pigments. This happens because light plays a crucial role in the production of pigments in animals, and without it the pigments either break down or aren't produced at all.

A spikey harlequin ladybird larva found sheltering underneath a leaf.

The trees in the woodland were humming with insects: hoverflies scooting left to right, wasps of various sorts buzzing all around, ants marching up the tree stems and beetle larvae in the moss-covered logs. Indeed, trees and their leaf litter support thousands of invertebrate species. Even decaying wood supports more than 2,000 different invertebrate species in Britain, which depend on it to complete their lifecycles.[5] In the UK, 1,178 invertebrates are known to be associated with oak trees alone.[6] Of these species, 320 are only found on oak and 229 species are rarely found on species other than oak – that's 549 invertebrate species that largely depend on a single tree genus.[7]

In tropical forests, this number is dwarfed by several orders of magnitude. Researchers estimate that as many as 30 million arthropod species call rainforests their home.[8] The late evolutionary biologist E.O. Wilson once found 43 species of ants on a single tree in Peru.[9] Another study found 82 species of ants on one tree in Brazil.[10] That's nearly double all the ant species in all the ecosystems of the UK. The Amazon rainforest alone is home to 10% of all the planet's species, and it's widely said that one square mile of rainforest might support more than 50,000 insect species – although I'm unable to verify this.

I continue walking through the woodland and notice a range of compressions in the vegetation, circular-shaped holes in the scrub and lines of eroded vegetation and soil stretching far and wide across the land like a vast network of secret pathways. They are indeed pathways, created by mammals: foxes, badgers, rabbits, voles, mice and other furry creatures as they commute, forage and search for mates. I notice elder trees in the distance – a sure sign of badgers being present as they eat the berries and germinate new trees in the woodland. Lo and behold, I pass a group of tiny holes in the ground full of animal poo – a badger latrine site. I continue walking, and 200 m away, a large mound covered in arch-shaped and badger-sized holes appears. It's an active badger sett: I can tell by the clear paw prints in the soil, the fresh spoil heap and the snuffle pits where badgers dig with their snout and claws to find nutritious plant tubers.

Our global forests are home to millions of mammals, from large elephants and jaguars to tiny rodents and bats. I've mentioned before about climbing trees to check for bats. They're protected by law in the UK as numbers are dwindling, largely owing to habitat destruction. To do a bat tree survey, you first need to look into the canopy from the ground and identify suitable features for roosting bats. These include holes, crevices and gaps behind bark that they like to sleep in. Most of the time, when I climb a tree, the holes are vacant. But occasionally, I gently place my endoscope in a hole and see one, two or sometimes many of our furry friends huddled together asleep before a busy night of foraging for insects. They're often found high up in the canopy, but they also roost lower down. I've found them in chest-height crevices, and on one occasion I even found a bat nestled within a crack of a wooden totem pole at a theme park. The crevices in trees serve as invaluable habitat for bats, birds and other creatures.

But it's not just inland that forests support wildlife. On the coastline, mangrove forests act as nurseries, supporting thousands of baby fish, crustaceans and other creatures. A mangrove tree can be compared to a fortress, standing strong against the powerful forces of the sea, its roots protecting and shielding the delicate ecosystems that thrive within its boundaries. And like the towers and battlements of a fortress, the tree's branches and leaves provide shelter and sustenance for countless creatures that depend on it for survival.

Forests support a vast number of animal species. As such, their ecological importance cannot be overstated. Yet what about plants and fungi that live on and among the trees? I've briefly mentioned the marvellous mosses. There are around 1,000 species of bryophytes (mosses, liverworts and hornworts) in the UK – many of which live in damp woodlands – and 20,000 worldwide.[11] These bryophytes, along with fungi, ferns, algae and lichens, are collectively known as cryptogams, and they are vital to the ecological integrity of forest ecosystems.

Cryptogams

Cryptogams, like microbes, are under-valued organisms. Despite sounding like blockchain technology, the term 'cryptogam' refers to plants and plant-like organisms that multiply by producing near-invisible spores. They don't produce flowers or seeds, which you can typically see with the naked eye. And that's where the name comes from – crypto (or Greek 'kryptos') means hidden, and gam or gamete means *reproductive cell* (actually coming from the Greek for wife), so cryptogam translates to 'hidden reproductive cells'.

In 2012, the Max Planck Institute of Chemistry in Germany calculated that the cryptogams covering the planet's soils, rocks and trees – and

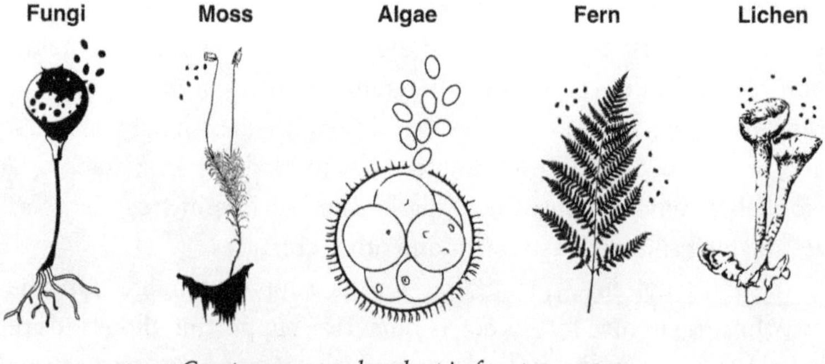

Cryptogams are abundant in forest ecosystems.

more recently, building surfaces – sequester around 4 billion tonnes of CO_2 per year, corresponding to more CO_2 than is emitted by all the world's cars combined.[12] Yet we rarely hear about cryptogams. The Max Planck scientists also suggested that cryptogams take up around 49 million tonnes of atmospheric nitrogen (N_2) annually.[13] This means cryptogams account for approximately 50% of the biological nitrogen fixation on land.

Lichens are intriguing cryptogams – a form of nested symbiosis. They're a conglomerate of fungi (mycobionts) and algae and/or cyanobacteria (photobionts) working together with a core symbiotic microbiome. They cling to surfaces such as tree bark and rocks and play important roles in ecosystems. They largely go unnoticed by humans in the forest ecosystem, yet to many urban dwellers, they're an unjustified nuisance: sadly, a lichen's destiny is often to be jet-blasted into oblivion. Around 20,000 known lichen 'species' exist, and they often have some beautiful features; they're like tiny open jewellery boxes of complexity.[14] They provide food and refuge for many minuscule creatures in a forest.

Mosses are equally valuable (intrinsically and instrumentally) but equally demonised as nuisance organisms. If you type 'how to remove moss', 'how to get rid of moss' and 'how to kill moss' into a well-known internet search engine, you'll probably receive somewhere in the region of 654,000, 122,000, and 24,000 hits, respectively. Our obsession with scraping these miniature ecosystems away in urban areas is astounding. Forests provide refuge for mosses – some of our most ancient plants. Over 300 million years ago, vast areas of the Earth's surface were covered in luscious moss carpets, pumping out oxygen. Along with cyanobacteria, we have the humble moss to thank for much of the oxygen we breathe today. Some are known to survive temperatures of 72°C (potentially 100°C when dried out) and –272°C.[15] They stabilise soil, retain water to help other organisms thrive and provide valuable shelter for visible and invisible biodiversity.

Like mosses, algae are autotrophs – they generate their energy via photosynthesis. They're typically found in aqueous environments such as lakes, rivers and oceans. But algae also live in the upper layer of soil throughout woodlands. Most soil algae act as cementing agents that bind particles together (called 'aggregates'), thereby preventing soil erosion. They can liberate large quantities of oxygen through photosynthesis in the soil, thereby facilitating soil aeration. Algae also maintain soil fertility by transforming sunlight and CO_2 into sugars, which insects, tardigrades and crustaceans in the forest ecosystem feed upon.

Ferns and closely related groups are also cryptogams. Ferns can produce millions of spores. Each dispersed spore, often aided by the wind, can potentially give rise to another plant. They appeared millions of years before dinosaurs and still thrive in various environments today, especially in forest ecosystems. Ferns can grow to a colossal size, examples being the towering green cathedrals mentioned in Chapter 1, but tiny ferns also exist: *Azolla cristata* is a water fern with 5 mm fronds – about the size of a drawing pin.[16] Ferns provide vital microhabitats for small animals in woodlands and rocky places.

Unlike the other cryptogams, which are autotrophs, fungi are heterotrophs, meaning they acquire their energy by feeding on organic matter. Three main groups of fungi exist (although these can be subdivided into many others), and they're abundant in forest ecosystems:

- multicellular filamentous moulds;
- macroscopic filamentous fungi that form large fruiting bodies, that is, mushrooms;
- single-celled microscopic yeasts.

Fungi make up some of the largest and smallest organisms on the planet, and scientists estimate that 12 million species exist.[17] They're the unsung heroes of trees and other plants. Indeed, mycorrhizal fungi form essential symbiotic relationships with soil bacteria and plant roots, allowing plants to draw in water and nutrients. In return, the plant roots shuttle carbon, sugars and other beneficial 'exudates' to the fungi. The structures and colours of fungi can be stunning; their shapes are like works of art. The more you look, the more you feel part of a truly unique world.

So forests are home to diverse plants and fungi, as well as animals of all shapes and sizes. There are 70,000 tree species worldwide and over 300,000 species of other plants.[18] A large proportion of these plants live among the trees in forests, and around 2,000 new plant species are discovered or described every year.[19] Beneath a forest's canopy, a kaleidoscope of understorey plants bloom, from delicate wildflowers to vibrant ferns. Our cryptogam friends carpet the forest floor and trees' bark while climbing vines and epiphytes cling to trunks and branches. Orchids, bromeliads and countless other botanical treasures form part of a community. Plants, together with microbes and fungi, play a crucial role in the forest ecosystem, creating a multitude of habitats that sustain a wide array of wildlife.

Oxygen

Trees and forest plants (like all other plants) release oxygen when they take energy from the sun and CO_2 and water from the environment. This process is crucial to photosynthesis, a remarkable chemical reaction that enables plants to produce glucose, their essential source of food and building blocks. It takes six molecules of CO_2 to produce a single glucose molecule via photosynthesis. Luckily for us and all the other aerobic creatures on the planet, six oxygen molecules are released as a by-product. A single mature deciduous tree, say a sycamore, produces around 100 kg of oxygen per year.[20]

Let's do a little thought experiment. It's been estimated that humans alone breathe in around 9,500 kg of air per person every year. There are currently 7.9 billion people on the planet, therefore humans alone collectively breathe in over 75 trillion tonnes of air each year. However, there's an important quirk we need to take into consideration: oxygen only makes up around 23% of the air by mass. Moreover, we only extract around a third of the oxygen from each breath. This means each of us extracts approximately 740 kg of oxygen per year. Collectively, this equates to humans extracting nearly 6 trillion tonnes of oxygen from the air each year. Suppose the average mature tree produces around 100 kg of oxygen each year. In that case, it works out at roughly seven to eight trees worth of oxygen required per person per year or close to 60 billion trees for the entire population. As mentioned earlier, much of the oxygen we breathe today came from cyanobacteria and mosses millions of years ago. But trees and other plants play a vital role too.

It has been estimated that around 130 billion wild mammals exist on the planet, which is only 4% of the total biomass of all animals, all requiring oxygen for cellular respiration.[21] There are also roughly 70 billion domesticated animals (e.g., cows, sheep, chickens) reared for the food industry each year, up to 60% of the total biomass of all animals on the planet.[22] Given the size range and respiratory variation among mammals, it's near impossible to calculate properly on the back of a beer mat! But, for a giggle, let's say, on average, other mammals require 300 kg of oxygen each year (that's 40% of the human requirement). That's another 60 trillion tonnes of oxygen extracted from the air each year, for a total of 135 trillion tonnes of oxygen.

Other organisms produce oxygen, including 50% of global oxygen released by blooms of tiny phytoplankton in the oceans, and the atmosphere was saturated with oxygen over evolutionary timescales.[23] Nonetheless, it gives an idea of how important trees and forest ecosystems are to the grand symposium of energy exchange and cellular respiration of animals on the planet. The act of inhaling the exhalations of trees unites us all.

Soil

Some consider soil to be the most valuable material on the planet. It's our – and many other creatures' – life-support system. Soil is the home of myriad microbes that decompose organic matter and fix nitrogen. It's also host to a plethora of other creatures, from worms to beetles, each playing an integral role in the ecosystem. Soil provides plants with an anchorage to grow and store nutrients and carbon. It filters rainwater and regulates waste. Soil is vital to all life on the land.

As a young child, I romped through our local fields in Derbyshire, oblivious to the mysterious world beneath my feet. Soil, the magical foundation of terrestrial life, was just a muddy medium to me back then. I dug mini-trenches and constructed mounds to ride bikes over, unknowingly disturbing the delicate system that sustains us all. I trampled on its forgiving surface, never appreciating its intricate web of living microbes and atoms from bygone organisms, silently nurturing the plants that surrounded me. It wasn't until I grew older and learned its ecological significance that I truly understood the treasure we often overlook. Now, with a new-found sense of awe, I cherish, study and protect the soil, for it's the heartbeat of our existence.

It can take around 1,000 years to produce 2.5 cm of topsoil, which is vital to biodiversity and human food security.[24] But we're currently losing topsoil at an unprecedented pace worldwide owing to human activity. Trees and forest ecosystems play a role in preventing soil degradation in several ways. First, they intercept rainfall, which is essential to prevent 'splash erosion'. The explosive impact of rain can break up soil particles, and trees help prevent this erosion. Tree roots help to bind the soil together – along with mycelial networks that interact with the roots – thereby enhancing the stability of the soil. They can also act as windbreaks, helping to prevent the wind from blowing soil particles away. Therefore, the wind-breaking capacity of trees plays an underappreciated role in preventing erosion.

Trees also regulate the water content of the soil via transpiration (the process of water movement through the tree and its evaporation from the leaves). It's a rhythmic release of moisture into the atmosphere. This plays a role in regulating the structure and nutrient content of the soil. The canopy of trees acts as a natural umbrella, shielding the soil from the scorching heat and helping regulate its temperature.

In addition to protecting soil, trees and forest ecosystems help to create the soil. It's all interconnected. Trees drop their leaves, fruits and flowers, helping to feed fresh organic matter into the soil. It's the same principles that underscore a compost heap. You keep adding your vegetable peels and other food waste to the heap, providing the microbes, beetles and worms with the food they need. In turn, these creatures break down the organic matter into a soil-like substance: compost. The trees do the same, feeding the tiny animals in the soil and the larger animals above ground; these regularly defecate and eventually die, thereby contributing to the soil's organic matter content, allowing new trees to grow – all part of the wonderful circle of life.

Carbon storage

Trees and their associated fungi are considered one of the planet's best carbon-capture technologies. The CO_2 captured during photosynthesis gets bound up into sugar, which is then used to build wood, roots and leaves. Plant cell walls are made of cellulose, essentially a long chain of glucose molecules (the sugar made during photosynthesis), so much of the carbon goes towards making cellulose.

Let's say, for the purpose of this discussion, that a fully grown deciduous tree reaches a height of approximately 12 metres. Now, here's the interesting part. This towering tree, if you were to weigh it, could clock in at a hefty 2 tonnes, accounting for its roots, branches and leaves. Every year it grows by about 5%, adding around 100 kg of wood to its sturdy frame. About 38% of that newly grown wood is carbon. The entire time a tree lives, it's pulling carbon out of the atmosphere and storing it in its tissues.

Forest ecosystems store carbon at different rates depending on the age and number of trees and the other plants and fungi present. Very young trees suck in carbon as if their lives depend on it (and they do), but the relative carbon storage is low. Once the trees grow larger and form a young woodland, they become excellently seasoned carbon capturers. Mature forests made up of middle-aged trees capture and store relatively greater amounts of carbon. The forest soil also stores vast amounts of carbon, but this depends on the forest type. For instance, the Arctic tundra (a mosaic of low-lying vegetation in often frozen soil) holds more carbon than the trees, whereas in a tropical rainforest where the nutrient turnover is rapid, the soil holds relatively little carbon compared with the trees.

Forest ecosystems are therefore carbon sinks. However, they are also carbon sources and emit tonnes of carbon into the atmosphere. For instance, when tree leaves fall to the ground, and trees die, much of the carbon is released during decomposition. The good news is that healthy and mature forest ecosystems sequester around twice as much CO_2 as they emit.[25]

Forests' dual capacity to sequester carbon and release it into the environment underscores their vital role in climate regulation. The various hydrological processes that forest ecosystems influence also mean they have a role in local, regional and continental rainfall and cooling. Indeed, a recent study found that converting agricultural land to European forests would likely enhance summer rains by an average of 7.6%.[26] And speaking of forest hydrology, let's go for a swim… high above the trees.

Flying rivers

Deep in the Amazon rainforest is the tallest structure in South America. It's a breathtakingly tall metal observation tower installed to evaluate an immense 'flying river'. This ethereal floating waterbody looks like a constantly swirling stream of dense fog.

Tree tissues act as a special network of highways for water molecules. As water travels into the tree roots and up into the tree canopy in a process called transpiration, it evaporates from the tree's aerial parts. Owing to the sheer size and density of trees in the Amazon rainforest, the water molecules form a river in the sky that's even bigger than the Amazon River itself. According to some reports, the trees in the Amazon release around 20 billion litres of water into the sky every single day.[27] You can imagine the impact that deforestation has on the flying river; but it also impacts other ecological processes in different parts of the world.

The Amazon's flying river flows to various parts of South America. As the primary airflow is to the south-west, the Andes mountains act as a giant natural barrier, redirecting vast amounts of water vapour to other countries such as Argentina and Paraguay. In 2009, a climate scientist said that huge areas of southern Brazil would be an arid desert without the flying river.[28] Recent droughts in Brazil have been attributed to the drying up of the flying river because of deforestation. Moreover, some evidence suggests that deforestation of the Amazon could reduce the rainfall in the USA.[29] This is known as a teleconnection – where a change or disturbance in one location can have significant effects or consequences in another location, often separated by great distances.

Meteorologist Eneas Salati showed that half of the rainfall over the Amazon rainforest came from the transpiration of the rainforest's trees.[30] In 2010, a hydrology student, Ruud van de Ent, created a global model to track atmospheric moisture and concluded that 40% of all the rain on the planet comes from the land and not the ocean as previously thought.[31] Science author Fred Pearce has reported that 70% of the rainfall in the Río de la Plata Basin in South America comes from the Amazon's flying river.[32] But it's not just the Amazon forest that has flying rivers. Ruud van de Ent's model suggested that China receives most of its rainfall from the boreal forests of Scandinavia and Russia, despite China being situated next to the largest ocean on the planet, the Pacific. It's a trans-Siberian flying river.

When the Amazon's flying river reaches the Andes mountains, it likely contributes to rock erosion, and the sediments are transported back into the Amazon River basin and eventually the ocean.

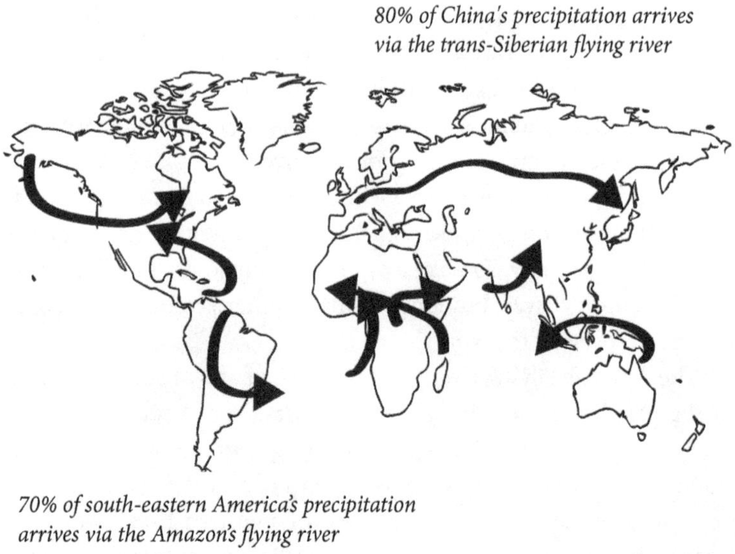

It has been speculated that once the sediment reaches the ocean it serves as a fertiliser for phytoplankton – microscopic algae and cyanobacteria. And as mentioned, oceanic phytoplankton produces around 50% of global oxygen. This is just one example of how the interconnections between trees and other global ecological processes run deep.

Flood protection

We often hear about the human costs of flooding. Some of us live in flood-risk zones and others see the images on the news of floods destroying houses. While certain ecosystems, such as wetlands, depend on seasonal floods to fuel their ecological processes, floods can also wreak havoc. This is especially true in the face of climate change, which intensifies and renders floods more erratic, compounded by inadequate land management practices.

Wildlife is significantly affected by flooding, with intense floods causing the destruction of habitats and resulting in animal drowning or displacement. A notable example is the 2012 Kaziranga floods in India, during which hundreds of animals, including endangered one-horned rhinos, perished.[33] Additionally, floods contribute to the erosion of riverbanks and substrates, carrying chemicals and disease agents that can harm downstream ecosystems. As discussed, flooding can be good for ecosystems in other scenarios – replenishing wetlands at the height of the dry season. Therefore, the planet needs flooding in small doses, and a diversity of habitats is favourable. Flooding plays an important part in maintaining this diversity. However, climate change means that the likelihood of extreme rain events is greater.

Trees play a crucial role in mitigating the impact of flooding by reducing its intensity. Forests are like giant umbrellas (albeit leaky ones). Through a process called interception, a forest's canopy slows down rainfall before it reaches the ground. The canopy acts as a shield, preventing raindrops from reaching the forest floor at high speeds. Instead, the leaves capture the water, which can then evaporate back into the atmosphere without directly affecting the ground. This interception mechanism helps regulate the flow of water, minimising the risk of flooding and allowing for a more balanced hydrological cycle.

The dense root system of trees and forest ecosystems facilitates water penetration into the soil, effectively storing it. Compared with nearby fields, the rate of water infiltration into the soil can be significantly higher, reaching up to 60 times. This enhanced infiltration capacity helps prevent surface run-off, reducing the risk of flooding. Research has shown that removing trees and hedgerows from rural areas leads to increased surface run-off, consequently raising the likelihood of floods.[34] The same thing happens in urban areas. If we continue to remove trees from our city streets, the risk of flooding will likely increase. Trees can also create a 'dragging' force and thus slow down the flow of water during flooding events. It's also worth mentioning mangrove forests again. In terms of flood protection, they are known to protect the landscape from perhaps the most acute of all flooding events – tsunamis. Mangroves can reduce the height and velocity of incoming waves and distribute the water among the mangrove canals that gave Venezuela its 'little Venice' name.

Earth systems science

When considering forest ecosystems, we can broaden our perspective to see Earth as a unified, intricate and adaptable meta-system. Our planet functions as a cohesive entity, shaped by myriad interactions between energy, matter and the diverse organisms that inhabit its ecosystems. Together, these components form the biosphere, an interconnected web of life. By recognising the interplay between forests and the larger biosphere, we gain a deeper understanding of how these ecosystems contribute to the functioning of Earth as a set of nested systems.

Although many Indigenous Peoples have recognised aspects of planetary systems for millennia, Western scientific disciplines have typically examined the components of the planet in isolation – taking a reductionist approach. In the past century, the emergence of Earth Systems Science has sought to foster a deeper comprehension of our planet.[35] This interdisciplinary field acknowledges that human activities, especially since the Industrial Revolution, have disrupted Earth's systems. As our understanding deepens, it becomes increasingly apparent that human actions have profound consequences for planetary systems. In response, the field of Planetary Health has arisen, recognising the intrinsic link between human health and the overall health of the planet. This interconnection highlights the crucial need to protect and preserve both human and planetary wellbeing in a mutually beneficial manner.

But how do forest ecosystems fit into this realm? As discussed earlier, trees can influence the hydrological and carbon cycles and, in doing so, influence planetary systems. Forest ecosystems support biodiversity and make up a considerable portion of the biosphere. And deforestation contributes to global climate change, as does the burning of trees that once lived millions of years ago (i.e., fossil fuels). Therefore, the influence of trees on the stability of the planetary system cannot be overstated.

Ultimately, everything is connected to everything else, and as my forest walk comes to an end, a flurry of organic linkages envelops my senses. It's good to break free of our brick-and-mortar cocoons and feel the raw threads of existence among the trees. They provide the air we breathe, the soil that feeds us and the beauty surrounding us. By 'us', I mean humans and more-than-humans alike. Yet they also represent a web of interdependence that we're only just beginning to understand. Trees have unrivalled ecological importance, and the fate of many species is intimately tied to them. Indeed, our fate is also tied to trees, and we must ensure they continue to thrive for generations to come.

Crayons – I Once Was a Tree

Many crayons are made from wax, and a significant portion of wax used in crayon production is derived from trees. Wax serves as the binder that holds the pigments together and allows them to be easily applied to paper. Wax provides a smooth and consistent texture, making crayons ideal for colouring and drawing.

The wax used in crayon production can come from various sources, including petroleum, but a notable portion is derived more directly from trees. One common source is the carnauba palm tree (*Copernicia prunifera*). Carnauba wax is obtained from the leaves of the tree, which is native to Brazil. It's known for its high melting point, glossy finish and excellent hardness, making it a popular choice for crayon production.

While tree-derived waxes are commonly used in crayons, other types of waxes, such as beeswax or paraffin wax, may also be used. Beeswax, as the name suggests, is produced by bees and offers an alternative to petroleum-based waxes. Non-carnauba waxes are also connected to trees, albeit indirectly. After all, trees provide homes for bees, and in apicultural settings they provide the wood for beehives. As mentioned before, trees contribute to the fossilised organic matter in petrol-based products such as paraffin wax.

PART II

RESTORE, REWILD, REGENERATE

CHAPTER 5

Forest Restoration

"Trees are sanctuaries. Whoever knows how to speak to them, whoever knows how to listen to them, can learn the truth."
—Herman Hesse

I'm now embarking on a local journey to the top of a limestone cliff in the Peak District, England. This is a place that indelibly etches a firm sense of solace within me. As I scale up the cliff, I carry unconventional cargo – a laptop, a flask of piping hot tea tucked away in my backpack and a rugged camping chair harnessed across my shoulders. This is all in aid of sitting peacefully in a patch of restored forest habitat and letting my fingers dance across the keyboard while drawing inspiration from the trees. It sounds cheesy, but I view it as a kind of method writing, akin to the immersive experience of method acting. With every key I press, I feel to be not an outsider looking in but at one with the perspective I seek to unravel.

As I reach the top of the cliff, I'm greeted by a vast sea of greater knapweed glancing down at me from the edge like floral guardians in a watchtower. They have pineapple-esque heads with bright pink and soft but spikey crowns – a firm candidate for the funkiest hairdo in the meadow. These guardians indicate a calcareous meadow growing on a bed of compressed and mineralised marine creature skeletons from bygone eras. The restored forest patch in this limestone habitat comprises semi-mature ash, wych elm, lime, oak and hazel. The ground flora is rich and includes a sprinkling of wild garlic and dog's mercury. This forest patch was replanted two decades ago and has undergone regular monitoring. The restoration project required active planting and nurturing of tree species characteristic of the target forest ecosystem. In addition, the area was protected to prevent livestock overgrazing.

However, not all forest restoration projects are the same. The options are manifold, from simply protecting remnant vegetation from the ravages of wildfires and overgrazing to more active and 'hands-on' approaches to restoring ecological functions. Natural regeneration is also an increasingly popular option – after all, mildly degraded forest ecosystems often need the space and time to grow by themselves. This is particularly true when humans behave in an unsustainable and exploitative way to the land.

Many forest restoration projects are now implemented under the umbrella of forest landscape restoration. This is the "planned process to regain ecological integrity and enhance human wellbeing in deforested or degraded landscapes".[1] Practitioners acknowledge that in addition to the functional ecological benefits (or those that sustain biodiversity), forest restoration is a facet of social equity and has important social and economic functions (or those that sustain human populations and their economic needs). This relatively new umbrella term encapsulates the need to meet nature conservation goals and the interests of the human communities that depend on the forests. It also promotes the need to care for the trees we plant rather than simply sticking them in the ground, ticking a box and walking away.

A striking example of the erosion of our duty of care for nature is the recent 'plastic nature' trend. We're buying synthetic grass, flowers and even animals at unprecedented rates. The global artificial grass market alone was valued at US$4.05 billion in 2021 and is expected to increase dramatically over the next decade.[2] I ran a basic back-of-an-envelope calculation based on current and projected revenues and the price of artificial grass per square metre to see how much of the planet we're covering in plastic turf. These rudimentary calculations suggest that by 2030 we'll be replacing natural habitats with around 60,000 ha of the synthetic stuff each year. That's nearly one and a half times the area of Barbados in artificial grass annually, and it probably has a rising trajectory. I've also seen plastic hedgerows installed by urban developers; and garden centres now sell a polymeric menagerie and nearly as many plastic trees and flowers as real ones, along with a sweetshop-like range of different potions to kill so-called pests. Caring for the earth is swiftly being replaced by convenience and the desire to be neat and tidy, yet the concept of caring is vital to forest restoration.

Richard St Barbe Baker

Richard St Barbe Baker OBE (1889–1982) was a global conservationist. Many would say he blazed a trail for contemporary forest conservation and restoration practitioners, and he was known as the 'man of the trees'. He began his mission to save global forests in 1922 when he started the first environmental non-governmental organisation (NGO), Men of the Trees, in Kenya. He once set a pledge for forest guardians: to "plant ten trees a year, take care of trees everywhere, and do one good deed every day".[3] Richard pioneered many of the concepts we talk about in forest restoration today and was ahead of many people in discussing sustainable development, agroforestry, rainforest conservation, biodiversity and Indigenous rights.

Richard developed his connection with forests from a young age. He travelled to the University of Saskatchewan in Canada from the UK to study Theology, to become a minister. While there, he met his Whitecap Dakota First Nations neighbours, who opened his mind to our deep interconnectedness with the natural world. However, strapped for cash, he went into the forestry industry to become a lumberjack – one of the main careers in Canada at the time. He quickly noticed the unsustainable practices rife in forestry and decided to go to England to study the science behind this discipline. After his studies, he joined the colonial service. Travelling to Africa, where he thought his job was 'conservator of forests', he discovered that his role was to assist in their exploitation. Probably influenced by his encounters with his Indigenous neighbours, Richard realised he had to protect the forests and gained deep spiritual convictions about the unity of life. He began to think about restoring forest ecosystems on a large scale in the 1920s, when most Westerners were still thinking about forests just as a resource. He developed a strong relationship with Indigenous communities in Africa and, according to author Paul Hanley, got into trouble for being "too sympathetic towards them".[4] On one occasion, when an English officer was going to strike an African worker, Richard blocked him, took the blow and had his collarbone broken. In addition to other incidents, this led to him being rejected from the colonial services.

Richard witnessed the effects of deforestation on the environment and on the diverse peoples across Africa. Widespread desertification was taking place, leading to the displacement of many communities.

This inspired Richard to set up Men of the Trees, initially aiming to encourage African communities to curtail their destruction of forests – to "develop a tree sense in every citizen and to encourage all to plant, protect and love trees everywhere".[5] In its heyday, the organisation was active in 108 countries.

Richard later developed theories about the history of the Sahara Desert, saying that it had once been covered in forests. He went on long trips across the desert to look for clues to back this up, and found signs of past forestation, including fossilised tree stumps. This led him to propose the Great Green Wall in the 1960s – an area of 11,662,500 ha of trees planted from the east to the west coast of Africa that would run through the Sahara Desert, and stop the expansion of the desert southwards.[6] Although the Great Green Wall didn't take shape during his lifetime, different nations are now coming together to create it. If completed, this will include billions of trees and be the largest human-curated, contiguously restored ecosystem on the planet. However, there are concerns that funding for the project continues to lag, while in similar projects women have been excluded from employment and local administrations have tried to privatise restored land that might instead be owned by local communities. Actions like these must be combated if the Great Green Wall is to succeed.

Richard St Barbe Baker inspired people to adopt sustainable forestry and restoration and is considered by many as the 'godfather of forest restoration', at least in the Western world. The Men of the Trees later morphed into the International Tree Foundation, which is still very active. It has an important motto: "When tree planting is done right, it transforms landscapes, communities and livelihoods."[7]

Community-led forest restoration in Oaxaca

Before the arrival of the Spanish conquistadors in the sixteenth century, the forests in the Coixtlahuaca district of Mexico supported more than 100,000 people. However, the forested hills have since succumbed to mass clearance and extreme erosion. As a result, the soil is now largely lifeless rock, and it strains the imagination to think of the vibrant forest ecosystem that once thrived here.

In pre-invasion times, the effervescent landscape bustled with species and had more than enough water to nourish many thousands of Mesoamerican Peoples. Yet the landscape now struggles to support fewer than 3,000 inhabitants. The land is largely devoid of vegetation, and it's a challenge to retain and collect the water needed to survive. However, all is not doom and gloom. Around two decades ago, the rural communities gathered to devise a plan to restore the degraded soil and revive the ecosystem. Unfortunately, organic foundations were absent. They had to start at the beginning, crushing the rock to retain water. This was extreme restoration.

Horacio Miguel, the mayor of Oaxaca state, remarked: "We aren't starting from zero, but rather from less than zero."[8] This meant the communities had to find trees and shrubs suitable for soil regeneration. They found a friend in Gregg's pine, a tree with a good reputation for growing rapidly in poor-quality soils. It's not an endemic species but grows further north in Mexico. Unfortunately, replanting native species is currently impossible because the soil is so degraded.

The decision to plant a non-endemic species came with many objections – although native species that were planted invariably died. Gregg's pine is currently being used as a 'nurse' species. The local ecologists know it can't reproduce in the area and will ultimately die, but it's helping to restore the soil rapidly and lay the foundations for other species. Other trees, such as white oaks, juniper and flint, are already beginning to sprout around the young pine trees. Squirrels, snakes and woodpeckers have made a comeback, and coyote populations are returning. It's crucial that wildlife is returning to the area. For one, the increased diversity provides ecological complexity and enhances forest resilience, but wildlife also plays another vital role in the restoration process. As mentioned earlier, when birds and mammals eat fruits and seeds, these eventually germinate in a rich natural fertiliser. The seeds are also spread far and wide, enhancing forest restoration by natural regeneration. In this extreme case, it would be difficult to restore the degraded ecosystem without providing a helping hand in the short term; but in due time, natural processes will assume more control, enabling the local community to actively participate in nourishing the ecosystem while deriving nourishment from it.

In the last two decades, the communities of Oaxaca have turned over 20,000 ha of degraded land into flourishing young forests.[9] That's more than 23 times the size of New York's Central Park. Long may this continue.

The East Usambara Landscape (Tanzania)

In 2004, an ambitious ten-year forest restoration project started in the coastal and sub-montane tropical forests of East Africa, the same forests in which I spent time in back in 2010. They are one of the most endangered biodiversity hotspots on the planet owing to rapid deforestation. The East Usambara landscape is one of the largest coastal and sub-montane tropical forests patches in the region.

Many endemic species call the East Usambara landscape their home, including the Usambara eagle owl, the endangered Usambara weaver (a small but striking bird that scrupulously curates or 'weaves' intricate hanging basket-like nests in the tree canopy) and the critically endangered long-billed tailorbird. The area comprises 35 villages where roughly 135,000 people live. However, historic and ongoing deforestation threaten the lives of many species, including humans: the communities there depend on the forest ecosystem for survival. In 2004, the Tanzania Forest Conservation Group teamed up with the World Wide Fund for Nature to start a forest landscape restoration project.[10] To take the pressure off the forest reserves, community activities were developed, such as brickmaking, thus reducing people's dependency on timber for building projects.

Between 2006 and 2012, forest clearance in the area plummeted by around 88%. Moreover, the two forest reserves in the region were connected by a new 97 ha natural corridor. It's hoped this will help alleviate the pressure on species by curtailing habitat fragmentation and increasing dispersal and gene flow. Agroforestry was developed by villagers, combining 'timber' trees with spice trees (cardamom, cinnamon and cloves) and maize. This has guided farming away from the monotony of single-crop cultivation and bolstered biodiversity, providing a diverse habitat – for both humans and wildlife alike – and connections to other forested regions.

Hundreds of tree nurseries were established in the villages, where the communities were fully involved in growing different species. Training was also provided. This equipped them with the tools and know-how to grow saplings and recognise the importance of growing and caring for trees instead of just planting them. The villagers also regained their autonomy, being able to oversee their forests without external meddling. The project exemplifies the need to balance both ecological and social objectives.

The Gondwana Link

A couple of decades ago, satellite aerial imagery revealed the extent of vegetation loss in Western Australia's biodiversity hotspot, an area of native vegetation in the south-western corner of the state. As elsewhere in Australia, ecosystems have been degraded since European settlers arrived in the early nineteenth century. Captivated by the striking satellite images that revealed the devastating loss of two-thirds of the native vegetation, a passionate collective of activists and ecologists were ignited with a fervent determination to embark upon an audacious forest restoration project: the Gondwana Link.[11]

The group realised that, despite the mass degradation, many patches of biodiverse habitat remained intact; but they were dotted across over 1,000 km. In these fragmented habitats a poignant truth has emerged: the remaining species confined within these isolated patches teeter on the brink of extinction. Denied the opportunity to mingle and intermingle with other populations, their genetic diversity and resilience are in grave peril, leaving them vulnerable to the whims of drastic environmental upheavals. If the patches are not reconnected, these remaining species could soon become extinct. Reconnecting the habitat patches is the primary aim of Gondwana Link. Notably, the restoration work required is also supporting the local Indigenous Peoples – the Noongar and Ngadju Peoples, who were dispossessed by European settlers. They're now regaining land rights and are co-leading restoration efforts.

One significant challenge has been land tenure as it relates to areas between the biodiverse forest patches. Much of this land was farmland, and Gondwana Link and its partners have purchased over 20,000 ha to enable the forest restoration. It's estimated that over 14,500 ha have been planted and restored since Gondwana Link's inception. Where once there were desolate stretches of farmland, now there are vibrant swathes of native species flourishing as they once did in the landscape's prime. It's a ballet of rebirth, as Gondwana Link's dedicated team works to revive the habitats one patch at a time. Importantly, the Noongar and Ngadju Peoples are reviving the 'cool fire' method mentioned earlier in the book. Fire management is desperately needed as vast areas burn uncontrollably, worsening damage to the sensitive ecosystems. Gondwana Link is the first project of its kind in Australia, and ecologist Richard Hobbs recently described it as "the most exciting thing happening in restoration in Australia, if not the world".[12]

In mid-2023, I spoke to Gondwana Link CEO Keith Bradby to learn more. Keith was instrumental in establishing the project back in 2002 and continues to lead it to this day. I asked him, "What is the current status of the Gondwana Link project, and how long will it be until the core goals are reached?" He replied, "The core goal is to restore 1,000 km of contiguous habitat. I feel optimistic about doing this by 2030." Keith went on to say that the team had secured more land in the last 15 months than during any other 15-month period in the project.

I asked what he meant by 'securing land'. He told me that part of the Gondwana Link model was to bring ethical investors and so-called carbon entities on board. These organisations invest in the restoration of the land, and in return they receive carbon credits and offsets. This offsetting is considered a way to make up for past misdeeds and to pave the way for a cleaner, greener future. Carbon credits are tradeable permits that represent a reduction in greenhouse gas emissions, particularly CO_2. Therefore, companies spilling out tonnes of carbon can buy these credits to offset their impact.

Carbon offsetting is controversial, as I discuss in Chapter 11. However, if it's done in the right way and by ethical investors, it can be beneficial for the environment. Importantly, the money generated from the credits often goes towards funding projects that help to reduce carbon emissions, like renewable energy or restoration and reforestation projects like Gondwana Link. Part of their model is to get investors to buy carbon credits so they can use the money to purchase degraded land. They then return the land to local Indigenous Peoples and pay them to help restore it. However, in my view, we should take carbon offset schemes with a pinch of salt – they may benefit the target ecosystem, but it's what they offset that's the issue. They often conceal a simple truth – if we want to stop emissions, we need to stop approving and subsidising things that cause emissions to go up.

Keith explained: "We secured 4,500 ha last year, and if everything goes as planned, we will have picked up another 2,500 ha this year. The difficulty is working in the market because land prices have more than trebled recently, and getting the right properties in the right place at the right time is tricky. We could theoretically buy the 1,000 km of land [2 km wide], but it's challenging because much of the land is covered in crops, which have become highly profitable. Furthermore, amidst the turmoil of the war in Ukraine, a worrying consequence has emerged in soaring crop prices." Indeed, in 2023, the prices of grains and beans skyrocketed by an astounding 60% or more in certain regions, triggering a worldwide food crisis that worsens an already dire predicament. Yet amid this chaos a paradox is arising as certain landowners seize the opportunity to reap hefty profits. As a result, acquiring land to restore the Gondwana Link becomes woven with complexities and challenges at every turn.

Keith told me that people see Gondwana Link as a symbol of hope in these challenging times, telling him on one occasion that it's Australia's biggest and most successful restoration effort. Keith replied, "Shit, is the situation that bad?!" This sense of shock reflects his reasonable assumption that such efforts should be universal.

I asked Keith what he had learned from being in the field. He rightly said, "I don't really view it as the field. It's more like something we're a part of, rather than visit to study." This struck me as a refreshing perspective that chimes strongly with the philosophy of many Indigenous Peoples. I asked what he'd learnt from Indigenous partnerships. He said, "It goes right back to the big stuff – my whole approach to country, life and knowledge. Such partnerships have reinforced my belief that we need to view ourselves as part of the land, not separate from it." He continued: "In terms of traditional ecological knowledge, I once sat with some Indigenous and non-Indigenous people. One of the non-Indigenous members of the group asked the Indigenous members, 'Do you still have that intimate knowledge of fire management? It's been a long time since your ancestors practised this on a regular basis.' The Indigenous members said, 'Well, it's not "knowledge" as you'd understand it. When we walk out into the bush, we draw upon personal experiences and intuitively know what to do.'"

Intuition is a complex and often misunderstood phenomenon. It refers to the ability to understand or know something without conscious reasoning or explicit analytical processes. It's based on close to unconscious processing (what the Nobel Prize-winning psychologist Daniel Kahneman calls 'System 1 thinking'), experiential learning or innate predispositions. There's a growing body of evidence from various fields supporting the existence and importance of intuition, and it's this type of knowledge that I think the Indigenous person was referring to. It's tacit knowledge, which is difficult to codify and teach in a traditional sense.

Keith also explained how some Indigenous people had practised horticulture for millennia. They shaped trees such as the Salmon gum tree at a young age. By tinkering with the tree's growth pattern, it developed into a multi-stemmed giant that would collect water many years later. As a result, these trees are known as water trees. It can take a century before the tree reaches its prime water storage age; this demonstrates immense foresight and altruism.

The Gondwana Link team has also applied forestry techniques to establish old-growth forests, which requires knowledge of planting density. For instance, depending on the rate of seeding and seasonal success, you can grow plenty of seedlings. They'll be tiny powerhouses for five years or so, and then they'll start crowding each other out; but if you provide the right spacing, they'll grow faster and at a more natural pace and with minimal competition.

Thanks to the combination of ancient knowledge and cutting-edge restoration and forestry techniques, the Gondwana Link thrives. The project has brought together the best of all worlds to create a habitat corridor that spans many thousands of hectares, connecting fragmented habitats and reviving ecosystems that were once on the brink of collapse.

Good vibrations: eavesdropping on forest soil

Most of my academic research focuses on restoring ecosystems and crafting strategies to enhance biodiversity and human health. Microbiota (the microscopic organisms in an environment) often play a central role in my research. However, in another life, I was a consultant ecologist, a journey I began in 2011. Since then, I've spent considerable time working on molecular ecology and 'ecoacoustics' applications to study and monitor the elusive denizens of our natural world. These methods include collecting stray pieces of animal DNA floating around in the environment and recording the diverse sounds of nature. For instance, you can take a quick pond sample to sequence the DNA left behind by amphibian skin cells – indicating their recent presence. And you can leave a tiny unit (about the size of a credit card) hanging on a tree to record the ultrasonic sounds of bats as they swoop around and mop up thousands of winged insects each night. The latter application – investigating natural sounds and their relationships with the environment – has recently evolved into the scientific field of *ecoacoustics*, sometimes called acoustic ecology.

All organisms generate sounds of some variety. Many bats, cetaceans (porpoises, dolphins, whales) and some shrews can explore their environments three dimensionally using echolocation, which is a form of biological sonar. They can emit pulses of high-frequency sounds, which allows them to move around in the dark of night to navigate, forage, identify friends and enemies, and avoid obstacles. These sounds are often emitted at frequencies too high for humans to hear, also known as ultrasonic. Many birds, in contrast, vocalise to communicate through songs and calls. Typically, songs are long and complex, whereas calls are short and simple. Birds sing from their syrinx, which is like a double voice box at the base of their windpipe. Most of the time, bird songs and calls are within the audible range for humans. The field of ecoacoustics can help us learn a great deal about bird behaviour and help us protect their habitat, which often includes forests.

Moving onto other winged creatures, sometimes it's impossible to escape the wonderful buzzing sounds that insects make as they rub their wings and legs together and contract and vibrate their rib muscles. But animals (and plants) also make sounds simply by moving through their environment or expanding or breaking their body's cells.

In 2012, plant scientist Monica Gagliano observed: "The ability to use sound is not a prerogative of animals." She hypothesised that "it would be particularly advantageous for plants to learn about the surrounding environment using sound, as acoustic signals propagate rapidly and with minimal energetic or fitness costs".[13] Plants can respond to sound waves or vibrations in their environment. For example, a startling 20,000 known plant species use 'buzz pollination'. This is where the pollen is released from flowers only when they are vibrated at the correct frequency. Bees have coevolved with plants to vibrate their flight muscles at the appropriate frequencies to stimulate pollen release. Plants can also detect acoustic vibrations that change their behaviour and emit sounds that indicate stress.[14]

Ecoacoustics researchers can explore the vibes of the underwater world as fish and other creatures emit their own sounds, detectable using hydrophones (submersible microphones). Even the air bubbles released from the benthos after decomposition create acoustic signals.

In the subterranean habitat of soil lives a plethora of micro- (invisible), meso- (tiny) and macro- (relatively large) organisms, all creating vibrations as their cells divide and as they navigate through the soil to commute, hunt and mate. These creatures include microbes, worms, beetles and other creatures that form wonderfully diverse communities. The sheer number of underground denizens is staggering: it was recently estimated that 59% (and even up to 74%) of all species on Earth live in the soil.[15] Earthworms make sounds through motion and friction with soil particles and can detect vibrations. They form burrows through the soil as they seek carbon-rich areas, and these serve as preferential networking pathways for plant root growth, water flow and gas transport, all of which contribute to the soil soundscape. We live in a euphonious theatre of life.

'Worm charming' has been practised by humans for many years. This involves driving a wooden stake into the ground and rubbing the top with a piece of metal to create a grunting vibration. When the worms detect this, they rise to the surface. Gulls do something similar by stomping on the ground to catch an easy meal. Charles Darwin was familiar with the technique, remarking, "It is often said that if the ground is beaten or otherwise made to tremble, worms will believe that they are pursued by a mole and leave their burrows."[16]

Beetle larvae make sounds in the soil as they move around and eat, but they also make sounds as they stridulate, that is, rub their body parts together to communicate. The wonderful thing about stridulation is that it appears to be species-specific, much like the vocalisations of bats and birds. Ants also produce vibrations in underground nests, particularly when in trouble, and a fleet of rescue ants can hear these sounds and rush to protect their relatives from harm. Eavesdropping on this subterranean cacophony is now providing a meaningful, inexpensive and non-invasive way of understanding who lives in the soil community. A couple of years ago, I thought, "Well, if we can detect who lives in the soil using ecoacoustics methods, can we also use these methods to monitor forest ecosystem restoration?"

As discussed, restoration is the recovery of ecosystems that have been degraded or destroyed, most often by human activities. Restoring soils is vital to the rest of the ecosystem, as many of the organisms and processes in the soil are foundational to all life. Armed with these ideas, I contacted an ecoacoustics specialist in the UK, Carlos Abrahams, who studies the acoustics of birds and freshwater ecosystems. I worked with Carlos for several years before embarking on my PhD: we were colleagues at a consultancy where I received ecoacoustics training earlier in my career. We used ecoacoustics to survey and monitor bats and cetacean populations in a conservation biology context. I've since used these methods to monitor birds and have also studied other aspects of sound.[17] However, monitoring soil life was an entirely new rabbit hole. Carlos was interested in the idea of using ecoacoustics to this end following the restoration of a forest ecosystem, so we started a research project. We also collaborated with Martin Breed – a restoration scientist who brought another dimension of knowledge to the table.

We identified a study site in South Yorkshire, where a large forest is currently being restored. The site comprises the same habitat with several age classifications – also known as a restoration chronosequence. This means there are blocks of woodland that are 0–10 years old, 11–30 years, 31–50 years and so on. This was perfect because we could use these different age classes to monitor soil biodiversity and its response to restoration across time. We did some pilot work to test various sensors and recording devices in soil before conducting a full study in the summer, when a greater array of soil dwellers are in their active phase. Carlos and I would stand in utter silence to record the sounds, one of us using a sensor to record soil and the other recording a 'blank' above ground to ensure we fully understood which ambient noises might confound our soil results. We would let out a long and loud gasp of relief between each recording, overcompensating for the previous several minutes of silence.

We plunged the sensors into the ground, some in plots freshly stripped of their forest cover and others in forest areas replanted at least three decades before. We repeated this many times, counted the soil invertebrates by hand and analysed the sound recordings using special software to generate acoustic metrics. This allowed us to quantitatively characterise the sound within each soil sample. Unfortunately, many of the soil invertebrates were unidentifiable in the field (neither Carlos nor myself is an entomologist), and Carlos's expert colleague, 'Dragonfly Dave', was busy at the time. So we had to take photographs of each invertebrate and identify them later. As anticipated, we discerned a notably enhanced diversity of sound in the rejuvenated soils, a finding that was in strong correlation with the abundance of invertebrates dwelling in the soil, particularly earthworms. Earthworms emit sound from their bodies as they move, creating vast networks of tiny underground channels. These channels act like flumes, carrying water through the soil, and this produces sound. Other organisms likely use these channels to get from A to B, and fungal spores are transported by the earthworms, which interact with plant roots. Earthworms also digest organic matter, and soil particles fall into the channels. These activities all contribute to the subterranean soundscape – a hidden orchestra of life. And it's all happening beneath our feet, and usually unbeknown to our ears. But with this technology, we can tune into the sounds of an ecosystem as it heals. We published our findings in the journal *Restoration Ecology*.[18] So it turns out that Mother Earth sings both above and below ground, and we can learn from her song. The next step is to improve our understanding of the cacophony of nature's vibrations to help monitor the recovery of our forest ecosystems.

Testing an ecoacoustics sensor in a bucket of forest soil.

The study was picked up by *New Scientist* and *Mongabay*, which both included a short piece about it. This triggered an unexpected level of interest with surprising and often strange outcomes. I was emailed by the organisers of a large conference, ChangeNow, held in Paris. This was billed as one of the world's biggest gatherings, bringing together the most innovative solutions and changemakers tackling our planet's biggest challenges. The organisers included a Soil Hub installation to showcase the importance of soil and the myriad reasons to protect it, and they asked if they could use some sound recordings from our study to show the difference between restored soils and degraded or 'dead' soils. The sounds of the soil were listened to by thousands of people over three days.

Perhaps even more surprising and pleasantly bizarre was an email that I received from the Norwegian author of the book *The Secret Life of the City*, Hanna Hagen Bjørgaas.[19] Hanna asked if I had any sound files that she could use in an orchestral piece to be composed by world-renowned composer Kristine Tjøgersen, with which she was collaborating. I explained that earthworm movement and digestion in the soil were the most likely cause of the bubbly sounds behind the white noise in the file I sent. "The bubbling sounds are amazing! Kristine (the composer) loved them as well," Hanna replied. "She will ask the orchestra to play the sounds as accurately as possible. I will keep you updated. The piece will be played in Oslo in September, as well as several places in Germany and Northern Norway." *Night Lives / Nattliv* was indeed performed later in 2023.

At this point, I thought about how strange but beautiful life can be. An earthworm wriggling around in the soil beneath a restored forest in South Yorkshire, was now being listened to by thousands of people at a conference in Paris and by theatregoers attending performances by an orchestra throughout Europe. I realised that this study had the potential to raise awareness about the significance of living soils in forest ecosystems.

Rise of the drones

During my PhD, I was awarded a small grant for a research project using drones, or unmanned aerial vehicles. The idea was to fly one around trees to take thousands of images. These images would then be stitched together to create a 3D model of the trees. I'd then collect microbiome samples from across the trees and use special software to integrate the microbiome data with the 3D models. This would allow people to view how the microbiome was distributed across the tree and how it changed over time. The time element added another dimension; we called the project '4D microbial cartography'. A tree's health depends on its microbiome. This is much as in humans, who depend on the microbiome of the gut, skin and airways to protect them from diseases. Being able to view the tree's microbiome and understand its dynamics can contribute to understanding the tree's health. While this study is ongoing, it highlights just one example of how drones can be used. I've used a drone to create habitat maps, monitor elusive species and create 3D models of trees. They're useful, particularly from a biodiversity monitoring perspective.

3D model of a tree using a drone – the detail of the branches was quite impressive.

My experiences led me to consider whether people were using drones for other applications in restoration ecology. After giving a talk on the subject at the World Conference for the Society of Ecological Restoration in Cape Town, South Africa, I met with colleagues Martin Breed and Peter Harrison, and we decided to write a review paper on the existing and emerging uses of drones in restoration ecology.[20] We found that drones can be used for various purposes at each stage of a forest restoration project: these can be divided into planning, implementation, and monitoring.

Forest restoration planning phase

Envision a drone soaring through the skies, collecting images and creating a mosaic depicting the sprawling vegetation that blankets the land. This can provide ecologists with a trove of valuable baseline information, allowing them to unravel the intricate story of the landscape before them. Drones are able to reveal the extent of degradation that has befallen the terrain, as well as the density of forests and the height and age of towering trees. This information can help to develop the precise interventions required to breathe life back into the land. Drones also possess a remarkable ability to assess plant health. With their advanced sensors, they can peer into the intricate nuances of leaves, unveiling subtle signs of stress. Armed with this knowledge, ecologists can discern plant health patterns, allowing for targeted interventions and nurturing the rejuvenation of fragile ecosystems. For example, multispectral cameras can be attached to drones, enabling estimations of how much chlorophyll trees contain – this is the 'green' pigment in the leaves that enable trees to photosynthesise. And drones can be used to count wildlife. This can be important to understand the baseline communities of animals in the forest or deforested landscape, allowing ecologists to plan an intervention to improve ecosystem health.

Forest restoration implementation phase

In the implementation phase, drones can disperse tree seeds and fertiliser across the landscape.[21] A drone is mounted with a special seeding device. Like miniature torpedoes of hope, they can unleash a cascade of seeds with the aim of bestowing life upon the barren, degraded lands below. The seed packages hold promises. They cradle the potential of growth as they are infused with fertilisers, nurturing the nascent seeds and coaxing them to germinate and flourish. However, the survival rates of trees are currently much higher if the seeds or saplings are planted and cared for by people on the ground. Drone-based seeding still requires considerable improvements.

A novel application of drones in forest restoration is wildfire tracking and control. Uncontrolled wildfires can devastate forest ecosystems. A fire-fighting drone can be mounted with a thermal camera to spot small wildfires. Another contains ecofriendly fire-extinguishing balls that it can drop on the fire, while a third acts as a communication hub, allowing the fire scouting and fire extinguishing drones to talk to each other across long distances.[22] It should be noted, though, that in the right conditions fires can benefit ecosystems. In this instance, there's the potential to use drones to start controlled fires. Indeed, researchers have recently developed fire-starting pellets that can be dropped from drones.

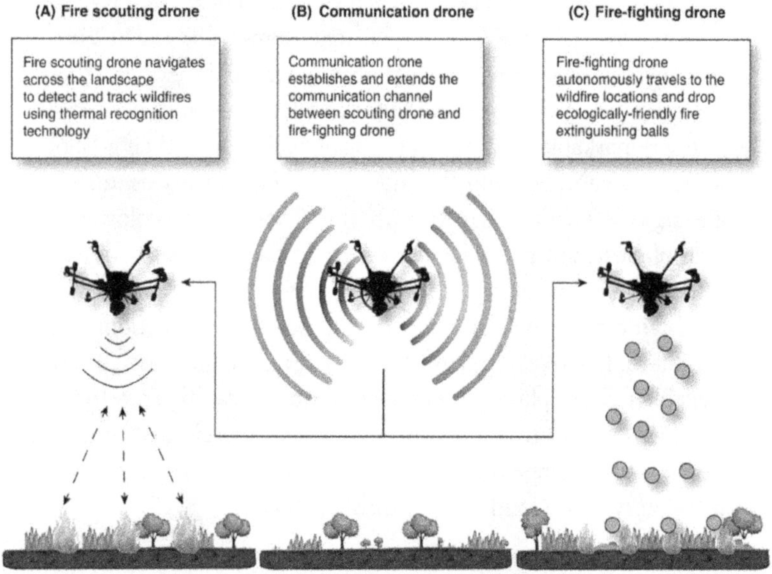

Wildfire-fighting drones.

Forest restoration monitoring phase

In a monitoring scenario, drones are able to evaluate whether a forest-restoration project's goals are being met. They can inform management strategies and illuminate pathways to enhance the revival of ecosystems. As in the planning phase, drones can be used to map the habitat to check on progress or even to collect samples using innovative sampling devices. For instance, a soil sampling device (picture a mini-soil corer) can allow a remote pilot to fly the drone to a desired location and collect soil samples, offering glimpses into the health of forest soils. In addition, drones can collect elusive biological traces (such as DNA) from a diverse range of habitats. This can allow ecologists to paint a picture of animal presence without the need for intrusive trapping methods.

While drones have made an important contribution to ecological research and conservation efforts in recent times, there are significant drawbacks that cannot be ignored. First, the intrusive nature of drones can disturb wildlife if flight safety protocols are not strictly followed. In the wrong hands, drones can cause stress, behavioural changes and even potential abandonment of nests or habitats. Second, reliance on drone technology might undermine the importance of fieldwork (and I feel more averse to the term 'field' since reflecting on my enriching natter with Keith), as it eliminates the human element and connectedness between ecologists and the natural environment.

A rich history and future

The history of forest restoration is rich, as is its future. Many forest restoration projects should be grounded in social science, and the knowledge of Indigenous Peoples can enhance the healing of our planet. But traditional ecological knowledge must be respected, and Indigenous land rights, sovereignty and research leaders promoted. Technology can also be useful, from listening to the forest's soil to using drones to map and monitor restoration interventions. A multi-pronged approach is often needed, combining cutting-edge ecological sciences, traditional knowledge and modern technology. And in other instances, simply giving the landscape a helping hand and some space (and lack of interference) can allow nature to heal itself. Yet environmental stewards can enhance the natural regeneration of forests. After all, humans are very much a part of the ecosystem.

CHAPTER 6

Natural Regeneration

"The care of the Earth is our most ancient and most worthy, and after all, our most pleasing responsibility. To cherish what remains of it and to foster its renewal is our only hope."
—*Wendell Berry*

Imagine you're going about your daily life – let's say strolling down the road to buy some food from the local store. Suddenly, a huge shadow engulfs you and everything in your immediate surroundings. The source of this looming darkness is a giant human-eating griffin as large as an elephant swooping down from the heavens. The griffin is simply doing the same thing as you and looking for its next meal. But the bad news, for you at least, is that you're the top choice on this griffin's menu. As the shadow increases in size, your fight-or-flight response (the sympathetic nervous system) kicks in. You scurry around to find the nearest safe place, changing direction with the aim of confusing the looming predator. You dart one way, but the griffin isn't fooled. You dart in the other direction, but this griffin is on top of its game and manages to grab one of your arms with its powerful talons. Unfortunately, the giant griffin carries you away and enjoys its lunchtime meal.

Now let's rewind. Reimagine the moment when the griffin grabs your arm. In this reality, you have a back-up plan called self-amputation or autotomy. This allows you to detach the rest of your body from your arm, scurry away to a place of safety and regenerate your limb as if nothing ever happened. The griffin gets a light snack (your spasming arm), and you get to live another day.

As science fiction as this sounds, many organisms have evolved this self-amputation defensive strategy to help them escape predation. For instance, various invertebrates, lizards, geckos and salamanders can initiate autotomy. Let's take the iguana (a large lizard) as an example. The iguana's tail has fracture planes – deliberate points of weakness – across each vertebra. This allows a vertebral column to separate alongside the adjacent musculature, nerves, blood vessels and skin. Once the tail has separated, its nerves are still active. This often causes the tail to wriggle violently for a few minutes, thereby distracting the predator while the lizard scurries away to safety. The cells at the base of the lizard's tail begin to rapidly divide and grow, and after a couple of months, a brand-new tail is formed. It's thought that iguanas may have a high number of stem cells, which can turn into other types of body cells. This allows them to naturally regenerate their damaged (or missing) tissue.

This is a marvellous example of how a living entity can regenerate. Its body has years of complex cellular evolution on its side, allowing it to heal itself. By its nature, regeneration invariably follows a point or period of damage or disturbance. We can look closer to home for an example of this disturbance–regeneration cycle in nature: the human body. For instance, the human skin (our largest bodily organ) is continuously bombarded by stimuli (whether they be electromagnetic radiation, chemical pollution or particulate matter, friction or even noise) that disturb its biological equilibrium and thus its health. These can all degrade the skin's local environment. And this is important because our skin is the immune system's first line of defence: it protects us from harm. Therefore, our bodies have evolved to constantly renew the epidermis every few weeks. The old skin cells fall to the ground as we shed 5 billion of them each day.[1] In essence, we have a whole new layer of skin every month – our personal version of natural regeneration. The great evolutionary biologist Lynn Margulis once observed: "Every year, 98 percent of the atoms of your body are replaced."[2]

Another example of natural regeneration at the species level is a coppiced tree, say, a hazel. Coppicing is a traditional method of woodland management that exploits the capacity of many tree species to send out new shoots from their stumps once they've been cut back. This process is another great example of nature healing itself. Humans recognised the self-healing ability of certain trees thousands of years ago and began coppicing to acquire a timber resource from the same tree multiple times. The practice has important conservation value too, as it can extend a tree's lifetime. Moreover, certain animals have learnt to utilise the unique habitat that coppicing provides. For instance, the hazel dormouse feeds on the hazelnuts and builds a comfy nest in the middle of the fresh shoots – which provides structure and protection to this cute little rodent.

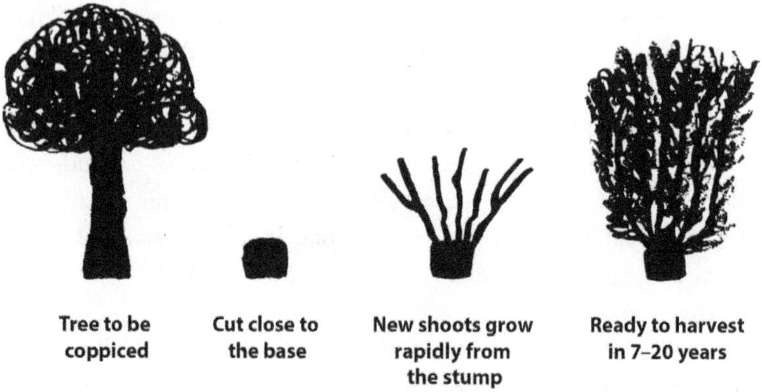

The coppicing cycle.

These are all examples of natural regeneration at the species level (iguana, human, hazel). But what about whole ecosystems, such as forests? Through the processes of self-healing, pollination, dispersal, decomposition and the almighty force of natural succession, entire ecosystems can regenerate. Given time (sometimes a very long time) and space, nature – from molecules and cells to species and ecosystems – can re-establish itself in the most degraded environments. In some ways, the natural regeneration of whole ecosystems, such as forests, can be compared to a wounded body healing itself. The body has mechanisms to repair and regenerate damaged tissues and cells. Similarly, a forest ecosystem can recover and regenerate from disturbances such as natural disasters or human activities. The body's healing process involves the recruitment of specialised cells and growth factors, and natural forest regeneration involves the growth of new seedlings, the colonisation of disturbed areas by pioneer species and the re-establishment of complex ecological interactions between different species. Both require time and a supportive environment to fully heal and recover. Much as the body can be strengthened through exercise and a healthy lifestyle, forest ecosystems can be strengthened and supported through sustainable management practices that promote natural regeneration and biodiversity.

While ecosystems can recover on their own, laying the foundations for this opportunity often requires dramatic changes in human behaviour. Our lives are so entwined with the rest of the natural world that the demarcation of 'protected' areas is sometimes not practical, and it can also be unethical in relation to local people. But when it's feasible, and when this process of natural regeneration is allowed to establish and develop, forests can once again become home to a rich tapestry of biodiversity and a bustling casket of ecological functions.

A recent study of tropical forests found that natural regeneration surpasses the more hands-on restoration methods in achieving tropical forest restoration for plants, birds and invertebrates.[3] This is the spontaneous recovery of native forests when tree seeds in the soil naturally germinate or when seeds from other areas colonise and establish themselves in the soils of degraded ecosystems. I liken it to the classic phoenix rising from the ashes. Just as the mythical bird is reborn from its own ashes, a forest can regenerate from the remnants of its own destruction.

Active or hands-on restoration methods often involve intensive propagation of trees in nurseries, followed by planting regimes, monitoring and after-care. Active restoration can often be costly but is important for the livelihoods of local communities. Nonetheless, there's a middle ground. Natural regeneration can be assisted through human intervention to, for example, protect regenerating forests from overgrazing, invasive species and fire, thereby efficiently restoring forest ecosystems at scale. This is called assisted natural regeneration. Essentially, it's giving nature a helping hand. In some ecosystems, assisted natural regeneration can take a long time, whereas in others it can be surprisingly rapid. The difference in the speed of recovery and the characteristics of ecosystems across the globe means that the carbon sequestration rate in global forests also differs: a recent study demonstrated that carbon storage during the first 30 years of natural regeneration can vary by 100-fold.[4] Assisted natural regeneration is best done in areas where favourable ecological conditions are already present and where local people are involved and benefit from the assistance. This means an evaluation is required to identify the most suitable areas – and where intensive tree planting is not necessary.

Adaptive potential

New generations of trees arise from the very soil that nourished their parents, the soil fed by their fallen ancestors. This soil is a matrix of minerals, microbes, organic matter and chemicals intertwined with the DNA and immense structures that emerge from the patient seeds. Naturally regenerated trees show adaptation to locally prevalent environmental conditions. Trees that sprout up naturally often survive better than planted trees. In natural conditions, new generations of trees are the offspring of those already thriving in the local area, making woodlands more resilient to diseases and a changing climate.

This frolic of life and renewal is no mere happenstance but a powerful adaptation that ensures the resilience and longevity of the forest. By propagating through natural means, trees evolve a diverse jewellery box of genetic material that enables them to adapt to changing conditions, from the ravages of pests to the stresses of a warming climate. Through this process of natural selection, the forest is strong and adaptable, and able to withstand the challenges that come its way.

Even in the face of devastating events such as wildfires and hurricanes, forests have the adaptive capacity to regenerate and recover, often with astonishing speed. Through a combination of specialised adaptations, such as resprouting, seed dispersal and symbiotic relationships with other organisms, forests can rebuild themselves from the ground up, creating new life and restoring its dynamic equilibrium. We should all tip our hats to the adaptive potential of forests and the resilience of life. The journey and transformation of the DNA in all the forest's organisms is biological poetry in action.

Successful natural regeneration case studies

Monks Wood, Cambridgeshire

Monks Wood is a National Nature Reserve – a place given protection for its important wildlife and natural features in England.[5] The wood is classed as an 'ancient woodland' (an area of forest with a continuous cover of trees since 1600). It was even recorded in the Domesday Book, published in 1086. Some may wonder what an ancient woodland has to do with natural regeneration. Well, a roughly 4 ha section of this woodland is an experimental woodland plot. To the untrained eye, it may seem as if the plot is part of the ancient woodland, but it has only been growing for sixty years.

Ecologist Richard Broughton, who works for the UK Centre for Ecology and Hydrology (UKCEH), told me about a note found in the UKCEH archive. Dating from the 1960s, it was written by Kenneth

Mellanby, the director of a forest research centre in Cambridgeshire at the time. The note describes the 4 ha plot before it was woodland: in 1960, the plot was an arable field containing a vast barley monoculture. The field had been deforested in the Roman period and cultivated or grazed ever since. But in 1961, the field was ploughed for a final time and abandoned. Mellanby's note reads: "It might be interesting to watch what happens to this area if man does not interfere. Will it become a wood again, how long will it take, which species will be in it?"[6] Mellanby's note planted the seed of an idea that developed into one of the UK's best examples of a naturally regenerated woodland: it was named the Monks Wood Wilderness Experiment. Richard described how the 60-year-old project began as a type of rewilding study decades before the term existed. It's a fine example of how giving the land time and space to naturally regenerate with a helping hand can lead to a thriving forest ecosystem that can help tackle the climate change and biodiversity crises. Richard told me how during the first 10 to 15 years a shrubland of thorny vegetation emerged, dominated by bramble and hawthorn, both favourites of berry-eating birds. The birds carried the seeds from one location and dropped them in another, spreading the thorny thickets far and wide across the plot. As the dense and spikey shrubland emerged, it provided an important physical barrier, protecting tree saplings from herbivores and wind. Wind-blown tree seeds began to take hold, particularly ash and field maple. The iconic English oak also began to grow, its nutritious acorns being carried and planted by Eurasian jays – who buried them with the intention (we presume) of saving them for a later meal. Indeed, over 50% of the trees in the Monks Wood Wilderness are thought to have been 'planted' by jays.

Jays: the ultimate restoration ecologists?

The jay is a relatively small corvid with a body length of around 35 cm. It has a pink/light rufous brown body plumage with a whitish throat and a prominent black moustache stripe. Perhaps its most striking feature is the prominent blue patch with black stripes on the upper surface of its wings. I personally think the jay is one of the UK's most beautiful animals, in appearance at least. Its alarm call is arguably not as beautiful, although beauty, of course, is in the eyes and ears of the beholder. It's safe to say that it sounds a bit like a screeching car. Indeed, its Latin name is *Garrulus glandarius*, meaning 'noisy acorn eater', and the old Celtic name for the bird is *schreachag choille*, meaning 'screamer of the woods'. Yet despite this head-splitter of an alarm call, the jay also has an impressive repertoire of more subtle calls, and it often mimics the sound of birds of prey such as goshawks and buzzards – an ingenious trick to keep said predators away or to scare off competing animals.

Another sign of the jay's ingenuity is its pastime of 'anting', meaning they pick up ectoparasites such as mites and ticks throughout the year. But how does the jay deal with this nuisance at best and potentially fatal risk at worst? The jay finds an ant nest – often those monumental wood ant hills you see in temperate woodlands – spreads its wings out over the nest and allows the resident ants to climb all over it. This bolshy behaviour elicits a defensive response in the ants, which react by squirting formic acid (the name comes from the Latin *Formica* for 'ant') all over the jay's wings. This acid just so happens to be a potent insecticide, which relieves the jay of its unwelcome hitchhikers. Some say it also helps condition the jay's feathers, a bit like using conditioner in your hair. But please don't apply formic acid to your hair, whatever you do: the outcome of this is bound to be unfavourable!

Along with eating insects and seeds, jays are voracious acorn eaters. Acorns are high in fat, carbohydrates, proteins and vitamins, and thus provide a rounded, high-energy meal. Individual jays can bury over 5,000 acorns in the autumn as a winter food stock, and despite being an intelligent species like other corvids, it's thought they simply can't remember where they cache all their acorns.[7] But in any case, they store far more than they actually need. This is crucial for an oak forest's regeneration, and indeed jays are credited with the rapid spread of oak trees across the UK after the last Ice Age. Many forgotten acorns grow into young saplings in the spring, and a lucky few become the great cathedrals of the forest that support hundreds of insect species, birds and other creatures. Owing to this interdependent symbiotic behaviour, one could argue that jays have a closer ecological connection to woodlands than any other corvids… perhaps any other animal.

Jays hop around on the ground, searching for acorns when they become ripe in early autumn. The immense intelligence of these birds allows them to find suitable hiding spots for the acorns while simultaneously searing particular landscape features into their memories, allowing them to return and feed during the winter months.

The Eurasian jay with its black moustache stripe.

Like other corvids, jays are associated with intellect, quick thinking, problem-solving, creativity and memory. I recall one day sitting in our garden in Hampshire, and a jay swooped in and perched on the rickety old-fashioned gazebo. It sat there looking at me with seemingly unwavering intent. I poured some seeds into a bowl and nudged the bowl towards the jay. The jay looked interested, at least from my (anthropomorphic) perspective, but appeared to be waiting for something else. I fetched a couple of acorns that had recently fallen from the oak tree growing closer to the house. I walked back and placed the acorns on the ground next to the jay. I noticed that one acorn had been infested by a knopper gall wasp, and so its appearance was deformed – the base of the acorn looked more like a star anise with gnarly festooned edges. I slowly walked away, and the jay swiftly moved in, poked the deformed acorn and brushed it aside. It then picked up the good-looking juicy acorn and flew off. It came back a few minutes later. I did the same thing, placing down a healthy-looking acorn and a deformed acorn, but also a healthy yet smaller acorn and a large but old and hollow acorn. The jay ignored the deformed, the smaller and large old acorn and went straight for the healthy large acorn. The jay was rapidly deciding which acorns were good or bad, even when multiple variations were presented. It also seemed like the jay was either expecting me to bring the acorns to it or, more likely, was simply waiting for the louder people at the other side of the garden to move away from the oak tree so it could collect its supper. Either way, it was a true joy to share that moment with the jay. Ultimately, this was an inter-species intellectual exchange and a mutually advantageous one. The jay received a meal, and I was gifted an awe-inspiring, off-the-scale feeling of nature connectedness. The jays' behaviour planted a figurative acorn of inquisitiveness in my mind. Do jays realise, on any level, that they're doing more than simply 'caching' acorns when they bury them each autumn? Are they intentionally planting the next generation of oak trees to preserve their livelihoods in the long term? I decided to take this question to Richard Broughton (mentioned earlier), who, among other things, is an ornithologist.

Richard replied, "I don't think we could ever really know. There's been a lot of research showing that jays are extremely intelligent and understand abstract concepts and delayed rewards, but I don't know if we could ever test that idea about oaks. The time between an acorn and a fruiting tree is beyond the lifespan of virtually any jay (around 25 years), so they would need multi-generational learning to pass down abstract knowledge about what an acorn can become. They don't have any first-hand experience of seeing an acorn become a tree that yields more acorns. Secondly, and more simply, they almost certainly bury acorns just to eat, not to plant them, so in a sense, the question could be, 'Does an oak tree know that jays will plant acorns, but not all will be recovered?' In other words, are the oak trees farming the jays rather than the other way around?! That's an even deeper question!"

"Oh my, now we're talking!" I replied. "Have oak trees co-evolved with jays, producing a particular variation of nut (acorn) that jays prefer because it increases the oak's future fitness and survival? I began writing about how jays are the ultimate forest restorers (but they likely don't know it), but now I'm wondering if the oak trees use the jays as their dispersers and planters, thereby restoring themselves! This resembles Richard Dawkins' extended phenotype, i.e., the genes that influence host behaviour tend to be maximised because of the behaviour itself, regardless of whether the genes are of host origin."[8]

A common example of an extended phenotype is a parasite's survival strategy. It can manipulate its host's behaviour to improve its chances of survival and reproduction. For instance, the lancet liver fluke infects ants and causes them to climb to the top of grass blades, where they are more likely to be eaten by grazing animals, the fluke's next host. The possibility of oaks evolving such capabilities to influence the jays' behaviour to benefit the oak would be an example of an oak's extended phenotype. In other words, the oak's genes might have effects beyond the physical boundaries of its 'body', such as in the environment or in the behaviour of the jay. This is a bit 'out there' but an intriguing prospect in the oak–jay–forest restoration debate, which I admit I'm currently only having with myself... and Richard.

I thought about it some more and went back to Richard once again. By now, he was probably thinking, crikey, what next? "I've been thinking about my original question, and one of the issues is that jays have no first-hand experience of seeing an acorn they bury turn into a mature tree that yields more acorns. Yet they *do* see oak trees of various ages, from saplings to mature trees yielding acorns, even if they're not the trees that arise from the acorns they bury . . . a stretch, though, I admit!"

I must reiterate that I think the most likely explanation (or at least the explanation with the most evidence) is the commonly cited notion that jays cache the acorns for food and sometimes forget where they are – and a lucky few grow into majestic oak trees. And jays have been observed to select the biggest, healthiest acorns, thus providing themselves with the most calories and nutrients. It just so happens that by selecting the biggest and healthiest acorns, the jays select those that are the most likely to survive and grow into large, healthy trees. However, despite this being the most likely story, I always feel compelled to question conventional explanations. Sometimes our cognitive biases force us to sleepwalk through life, placing a veil over the truth. And the argument for rejecting the hypothesis that jays may, at some level, understand that acorns grow into acorn-laden oak trees was not yet compelling enough to me.

Richard replied: "There are some big feats of intelligence involved in that. Firstly, the acorns that grow are probably from jays who've forgotten where they buried them or from jays who have died (unless they deliberately plant them to grow, but there's no evidence that any species apart from us does that kind of cultivation – there would be no reward for the individual as it would never live long enough to get any acorns back). Secondly, it would involve jays grasping the concept of things growing over time, which is pretty abstract and means they'd have a concept of 'deep time' beyond their own lives. That would be a huge leap. A jay might never link an acorn to an oak seedling because it would never know that an acorn planted by another jay was the thing that had resulted in the seedling and leaves that the other jay witnesses – it would never see and learn the whole process, unless, again, the same jay had the foresight to plant the acorn for the purpose of germination. The only way to bridge that is for them to see acorns and oaks at every stage of germination and growth and then make that link of things getting bigger, as you say. But that's a profound concept due to the timescale, as they would have to understand that the benefits would be for their descendants. So, they would be making a conscious altruistic sacrifice (don't eat the acorn today but provide acorns for future jays) and maybe even understand their own mortality and time beyond their lifespan. You might need to ask one of the corvid cognitive people at Cambridge – they're probably having conversations with jays!"

Richard's response was excellent, but it also got me thinking about evolution and another concept that Dawkins developed and popularised in the 1970s, that of the selfish gene.[9] This concept suggests that genes metaphorically strive for immortality. Through their role in reproduction and natural selection, genes can influence an organism's behaviour in a way that maximises their own replication, leaving individuals, families and species as mere vehicles for the genes' quest. In his book, Dawkins focused on behaviours we call 'altruistic'. This refers to behaviour that reduces the fitness of the actor but increases the fitness of the recipient, much as in Richard's point above. There are many examples of this so-called altruistic behaviour in nature. For instance, honeybees are known for it, as worker bees sacrifice their own reproductive potential to care for the queen and her offspring. Worker bees forage for nectar and pollen, tend to the brood and defend the hive, all at the cost of their own survival. Another example is that of prairie dogs. They live in colonies and have a complex communication system.[10] They've been observed sounding alarm calls to alert others in the colony of predators, even when it puts the alarm callers (themselves) at risk. Dawkins suggests that many of these behaviours are not at all altruistic but driven by the genes as the fundamental units of natural selection that act to preserve themselves, even if it means sacrificing the wellbeing of the organism that carries them. My point is that jays may not need to consciously understand that the benefits of 'planting' acorns would be for their descendants and thus altruistic. The process may be driven by genetics.

Therefore, what if jays bury plenty of acorns but don't collect all of them because some will secure nutritious nuts for future generations? Hypothetically, what would happen if they didn't do this? Let's say that the ancestors of jays and oak trees co-evolved with each other millions of years ago. Indeed, some of the extant genera of oaks evolved at a similar time to corvids after the K-T extinction event tens of millions of years ago.[11] Perhaps oak acorns evolved to biochemically attract jays in a symbiotic relationship. If the jays didn't leave some acorns to germinate and grow, the jay populations would no longer have the acorns that helped them survive and proliferate. This would provide the selection pressure needed for the jays to keep 'planting' the trees. Except that, as Richard (Broughton, not Dawkins) points out, individual jays may not acquire the knowledge that acorns grow into acorn-laden trees because a jay's lifespan is only up to 17 years, and most oak trees only start producing acorns when they are between 20 and 40 years of age.

What if jays were to see and learn about acorns growing into saplings and semi-mature trees, and see trees growing at different ages (some from acorns buried by previous generations)? And at some point, some jays deliberately – or accidentally – left acorns to germinate into saplings after recognising that more oak trees would grow, which led to a survival benefit.

I then started thinking about the jay's evolution in biogeography terms. Some think jays evolved in Asia and then spread across the European continent. Biogeographer Knud Jønsson thinks all corvids evolved in the area now known as New Guinea.[12] This means the possible symbiotic relationship between oaks and jays could have evolved on the Asian continent millennia ago. I started looking into Asian oak trees and came across the Chinese cork oak *Quercus variabilis*. This can start producing acorns after only eight years. Recall that the jay's lifespan is up to 17 years? Therefore, the planting behaviour and a jay seeing saplings grow into acorn-laden trees could easily have taken place during a jay's lifetime. The trait (also known as phenotype) could have been retained and carried forward across generations and geographic regions. Jays living today may simply be born with such tendencies through genetic means, replicating what their kin have always done for generations. This could be selfish (because evolutionarily, the individual jays may have benefited within their lifetime or, as mentioned earlier, their genes could be the driver), or it could be a form of altruism that's hardwired into many creatures that live in social groups or with strong family bonds.

But what if there was another reason that jays might deliberately 'cultivate' oak trees – not just for acorns? I recently found a passage in a book called *Wilding* by Isabella Tree that suggested jays might deliberately leave some acorns to grow because they understand that the young leaves (cotyledons) are important for infant nutrition.[13] Young jays apparently love to feed on these highly nutritious leaves when they first sprout into infant trees. If true, this would undoubtedly allow for a learned trait (or even a genetic tendency) to develop and secure oak tree generations. According to this anecdotal evidence, jays are aware that the acorns turn into young oak trees. Indeed, sometimes they'll return to newly growing saplings and feed the early leaves to their young. The saplings survive to grow into mature trees.

I decided to take Richard Broughton's advice and contact the bird research group at Cambridge. Alex Schnell has done extensive work on jay behaviour. She showed that jays, like some humans, can pass a variation of the famous Stanford marshmallow test.[14] This was a study in the 1970s on the human ability of delayed gratification. The Stanford researchers put a marshmallow in front of a group of children and told them they could have a second marshmallow if they could go 15 minutes without eating it. Whether the children were patient enough to double their payout is supposed to indicate willpower that will pay dividends later in life, for instance, at school and eventually at work. Passing this self-control test is purportedly a promising signal of future success. Alex found that jays can also pass a version of this test, and those who score higher on intelligence tests are most likely to have the greatest self-control.[15]

Alex kindly sent me a raft of papers that show compelling evidence to demonstrate that food-caching in jays is not a hardwired behaviour. Rather it involves some complex cognitive traits once thought to be unique to humans, such as episodic memory, self-control, future planning, desire attribution and perspective-taking. This suggests caching might be a learned behaviour in jays but that they can also plan for the future. The previously dominant hypothesis that only humans can anticipate and act towards the satisfaction of a future need – not cued by their present motivation – can be thrown out of the window.

To summarise, despite the best-supported notion that jays simply forget many of the cached acorns, which then grow into majestic oaks, there's anecdotal evidence to suggest that jays understand that acorns grow into larger and more complex forms. There's a theoretical basis in evolutionary frameworks, such as the genetic drivers of seemingly altruistic behaviours, but also the ability of jays to plan for the future. And don't forget – it might also be that oaks are restoring themselves and using jays as the vehicles.

Ultimately, it doesn't really matter whether jays realise they're planting the next generation of majestic oak forests. Their behaviour weaves together a tapestry of life that echoes through the ages by tipping a few grams of soil onto the humble acorn.

Other animal restorers and cultivators

My conversation with Richard led me to wonder about the lack of evidence that other species do this kind of cultivation. That we humans are the only cultivators of crops is a commonly accepted idea, but is it true? Like a literary treasure hunter, I started digging around the internet for scientific papers and stories about non-human animal 'farmers'. I started thinking about other animals that might be restoration ecologists, or at least plant cultivators that assist the natural regeneration process.

In my search for answers, I stumbled upon a peer-reviewed article entitled 'Root cropping by pocket gophers'.[16] This piqued my interest. The paper discussed how pocket gophers *Geomys* spp., solitary, root-eating rodents native to North and Central American grasslands, are presumed to acquire most of their food underground as they excavate a complex series of tunnels up to 160 m long. According to the researchers, burrowing takes 360–3,400 times more energy than surface walking, so clearly pocket gophers have very high energy requirements. The authors note that roots are often scarce at the depths of their tunnels (when gophers are absent), but they could supply up to 62% of their daily energy needs. The gophers create the conditions that favour root growth, spreading their own waste around as a fertiliser before harvesting or cropping roots. As Francis Putz, a coauthor of the study, remarked: "If farming requires that crops be planted, then gophers don't qualify. But this seems like a far too narrow definition for anyone with a more horticultural perspective in which crops are carefully managed – such as fruit trees in forests – but not necessarily planted."[17] The gophers' ability to scrupulously curate expansive networks of underground tunnels also shapes the landscape and the nutrient availability of the soil. They act like tiny tillers, aerating the soil. The aeration and nutrient turnover increase the productivity of plants. Therefore, the gophers are important players in restoring ecosystems, producing a complex mosaic of nutrients and soil conditions that stimulate plant diversity across the landscape.

The gopher, with its impressive pick-axe teeth.

The earliest-known example of cultivation, though, is thought to be that of the humble termite. Researchers have discovered the oldest known examples of 'fungus gardens' in 25-million-year-old fossil termite nests in East Africa.[18] They found that some termite species cultivate these 'gardens' in subterranean nests or chambers, helping to convert plant material into a more easily digestible food source. The origin of this symbiotic relationship between the termites and the fungus likely had a profound effect on nutrient fluxes across the African landscape, shaping the acacia scrubland. The evolutionary transition to such fungus agriculture by the termites increased the range of available habitats for both the termites and the fungi. Therefore, these subterranean creatures were likely involved in habitat creation on a massive scale – a gigantic feat for a tiny being.

Whether or not non-human organisms are aware that they are the ultimate ecosystem restorers, nature's artisans, it almost doesn't matter – much like the jays. A bustling conglomerate of invertebrates, birds, microbes, mammals, reptiles and amphibians contribute to pollination, seed dispersal, soil formation, tree health and fertilisation. And, of course, they are an intrinsic part of the tapestry of life that they ultimately restore through the wondrous process we call natural regeneration.

Back to Monk's Wood

In the intermediate shrubland stage of Monk's Wood, sunlight hit the ground, providing conditions for light-loving wildflowers to flourish. Herbivores were present in the local area, consuming tree saplings, but the thorny shrubs protected these young trees. As such, no tree guards were needed (this is a contentious issue, as we will discover in Chapter 12). As the trees rose towards the sky, their developing canopies eventually closed above the thorny shrubs, which became the woodland's understorey layer. The ground layer of herbs and wildflowers, the shrubby understorey, the accumulating deadwood and the towering canopy layer paint a picture of a structurally complex woodland that offers niches for a wide variety of life. But what was it like in reality? I had decided to pay a visit to Monks Wood to find out myself.

When I arrived at the wood, I was greeted by a plethora of butterflies. Their wings fluttered in the spring breeze and they danced like the stars of a graceful ballet, adding a touch of enchantment to the woodland glades. I looked down at the floor and noticed a rich purple hue all around me. As my eyes adjusted, I saw a carpet of bluebells stretching out before me, as far as I could see: a serene and tranquil sea of delicate blooms that invariably lifts the spirits. I put my ecologist's cap on, did a vegetation survey of the woods and was astounded by the sheer diversity of species. One of my favourites is silverweed *Potentilla anserina*. With delicate yellow petals that shimmer in the sunlight, this bloomed along the woodland edge. As the sun hit it, the silverweed took on a new radiance, its shiny leaves glowing with a soft and subtle light that illuminated the rugged track.

The diversity of bird species was also astonishing. It reminded me of being somewhere tropical but with temperate sounds. You may recall that this 4 ha woodland plot was only 60 years ago an arable field containing a vast barley monoculture. It exemplifies the idea that if we give nature space, time and nurturing, it will bounce back.

Costa Rica

Costa Rica, a small country nestled in the heart of Central America, has been a beacon of hope for environmentalists worldwide. Despite facing challenges such as deforestation, habitat loss and biodiversity decline, Costa Rica has made strides towards natural forest regeneration. Through years of dedicated conservation efforts and political will, the country has transformed itself into a world leader in reforestation.[19]

One of the key factors in Costa Rica's success story has been its strong commitment to sustainable development. The country has implemented policies that prioritise protecting natural resources and preserving biodiversity. It has also encouraged community participation in conservation efforts, empowering local communities to take ownership of their natural resources. As a result, Costa Rica now has 50% forest cover (much through restoration), providing habitat for countless species of flora and fauna.[20] This is noteworthy, given that in 1940 Costa Rica's forest cover was only 26%.

As the rainforests flourish again, the sounds of howler monkeys and the flutter of butterfly wings fill the air. The lush canopy provides a home for toucans, macaws and other exotic birds, while the forest floor teems with a myriad of creatures great and small, from jaguars and tapirs to tiny insects, fungi and other microbes. Indeed, Costa Rica is home to over 500,000 species of flora and fauna, many of which are found nowhere else on Earth.[21] This is actually around 5% of all the species on the planet, making it one of the most biodiverse places on Earth.[22] As a species, we're capable of destruction and creation, of tearing down the natural world. Yet in Costa Rica we can see the power of human action, the ability to heal and restore what has been lost, demonstrating our species' potential and capacity for good.

The Knepp rewilding project

In the heart of Sussex, there lies a place of rural wonder where the land is free to run its course. This is the Knepp rewilding project, a bold and visionary experiment in ecosystem restoration that has breathed new life into this modest patch of English landscape. Here, many of the tree canopies of this once degraded land are now thick with leaves and alive with the chatter of birds and the rustle of scurrying mammals.

As mentioned in the earlier chapters, for centuries the woodlands of Britain have been shaped by human intervention (which is not always a bad thing), their once-lush ecosystems tamed and transformed by centuries of unsustainable logging, hunting and agriculture. But at Knepp something different is now happening. Here, the forest and scrub are being allowed to grow and change as they will, without the constraints of urban development. The result is a woodland and grassland mosaic that feels both ancient and new. I visited the Knepp Estate when writing my first book, *Invisible Friends.*[23] As I walked among the gnarly oaks and rustling beeches, it was hard not to be reminded of the infinite possibilities that await us when we choose to have aspirations to live with the land.

There are plenty of pockets where you can lose yourself in the natural world and find yourself in its abiding rhythms. And we shouldn't divert our minds away from the fact that we need to allow nature to restore in a way that's harmonious with the needs of humans; after all, we're a part of the biosphere too! Through assisted natural regeneration and a cultural shift in mindset, we can work with nature to restore the planet. This chapter affirms life's natural and sometimes metaphorical will to endure. In the right conditions, with every sprout breaking through the hardened soil and each seed taking root, and after every season of falling matter, there's a rising power waiting on the horizon.

Paper – I Once Was a Tree

Paper has a rich history dating back over 2,000 years. The first known papermaking process was developed in ancient China during the Han dynasty (202 BCE–220 CE). The invention of paper revolutionised communication and paved the way for spreading knowledge throughout human civilisation. So trees have influenced language and knowledge throughout time. And through such development, trees have shaped our very thoughts.

It's estimated that one large tree can produce about 8,333 sheets of paper. This means one ream of 500 sheets of paper would require about $1/17^{th}$ of a tree. The paper industry is estimated to be the fourth largest contributor to greenhouse gas emissions among manufacturing industries.[3] However, it's also one of the largest renewable energy users, as many paper mills use the by-products of the pulping process to generate power. Recycling paper can help mitigate some of these impacts. It's estimated that recycling a tonne of paper can save about 17 trees, 7,000 gallons of water and 4,000 kWh of electricity compared with producing the same amount of paper from virgin materials.

Over 400 million tonnes of paper are produced each year. While digital alternatives have reduced paper usage in certain areas, paper is still widely used in various industries, including packaging and publishing.

CHAPTER 7

Chernobyl's Red Forest

"I can be changed by what happens to me, but I refuse to be reduced by it."
—Maya Angelou

On 26 April 1986, the Chernobyl nuclear power plant, located in the northern part of the Ukrainian Soviet Socialist Republic, experienced a catastrophic explosion that resulted in the worst nuclear disaster in history. The disaster occurred during a routine safety test of Reactor 4, which involved shutting down the reactor's cooling system and simulating a loss of power to test the emergency back-up systems. However, a combination of design flaws in the reactor, operator error and the absence of safety mechanisms led to a rapid and uncontrolled increase in reactor power. The resulting explosion and subsequent fire released a massive amount of radioactive material into the atmosphere, contaminating large areas of Ukraine, Belarus and Russia, as well as parts of Western Europe.

The immediate aftermath of the disaster was chaotic and confusing. Plant workers and emergency responders were exposed to high levels of radiation and suffered from acute radiation sickness. The Soviet government was slow to respond, initially downplaying the severity of the accident and failing to evacuate nearby towns and cities. It was not until two days after the explosion that the Soviet authorities finally admitted that an accident had occurred.

The consequences of the Chernobyl disaster were far-reaching and long-lasting. The immediate death toll was officially recorded as 31, although some estimates suggest that the actual number may have been much higher.[1] In addition, thousands of people were exposed to radiation and suffered (and continue to suffer) health problems as a result. The contaminated area around the plant, known as the Chernobyl Exclusion Zone, embodies humanity's greatest fears and darkest moments. It's a place where the boundaries between the natural and the unnatural blur, where the forces of destruction and creation collide – a nuclear storm of excited molecules and free radicals.

I can't help but imagine how the land bore witness to the abandonment of a once-thriving community. In its wake, the nuclear accident left a landscape of eerie desolation. Nonetheless, amid the ruins and shadows, life persists – trees, wolves, cranes and bees – illuminating the resilience of nature and the fragility of our existence.

It's safe to say that the Red Forest is one of the most contaminated places on Earth.[2] This is the area in the Chernobyl Exclusion Zone that received the highest doses of radiation from the nuclear disaster. The resulting clouds of smoke and dust, heavily polluted with radioactive contamination, smothered the once-thriving pine forest that carpeted this region. Prior to the disaster, the area was a dense coniferous forest with a scattering of oak and birch trees that was home to a diverse array of wildlife, such as deer, wild boar and lynx.

Unfortunately, when the nuclear reactor exploded, the trees soon died. The explosion contaminated the soil, water and atmosphere with radioactive material equivalent to that of 400 times the atomic bombings of Hiroshima during the Second World War.[3] The Red Forest's name doesn't come from the natural colour of its leaves but from the deadly radiation that caused its demise. This was caused by a combination of factors. When the reactor exploded, it released a massive amount of ionising radiation, which damaged the DNA of the trees and disrupted their ability to photosynthesise and produce chlorophyll. This is the pigment that gives plants their green colour, and without it the trees took on a reddish-brown hue. Additionally, the trees were exposed to high levels of radioactive particles that were carried by the wind and settled on their needles. These particles emitted energy that damaged the cells of the trees, leading to their death. In the post-disaster clean-up operations, most of the pine trees were bulldozed and buried in trenches.[4] These were then covered with a thick carpet of sand and planted with pine saplings. Some pine trees also started growing naturally, a sign that the Red Forest might one day become green again. And the good news is that despite all the destruction, the forests and their wildlife are once again thriving in this area, or so it seems. Whereas before the disaster the Exclusion Zone had

30% forest cover, now it's around 70%.[5] Up to 90% of the radiation from the Chernobyl disaster was trapped in the canopies of the conifers.[6] As the trees shed their needles, the needles scattered the radionuclides (radioactive particles) into the topsoil. In a few years, fresh trees started to absorb these radioactive elements. Researchers have found that the trees treated the caesium and strontium like nutrients since they are physically and chemically similar to potassium and calcium.[7]

Just like many other organisms (including humans), trees need potassium, calcium and other elements for various physiological functions. For instance, potassium regulates the opening and closing of stomata, tiny pores on plant leaves, facilitating water absorption and nutrients from the soil. It's also vital for activating enzymes that support photosynthesis. Calcium is a key component of tree cell walls, providing structural support, moving nutrients and regulating other metabolic processes. Deep within the cellular caverns of trees, their genes encode proteins that essentially tell them to acquire these elements from the soil via a swathe of chemical cascades. At this stage, they're on autopilot. A marvellously complex arrangement of processes occurs inside the trees' cells, triggering cell-to-cell communication and causing mini-transporters to move these elements from one tissue to another.

Trees also form mutual relationships with soil fungi known as mycorrhizae. These enhance nutrient uptake, including potassium and calcium, by extending their fungal hyphae (tiny filaments) around the roots and soil, increasing the tree roots' surface area and nutrient-absorbing capacity. Therefore, a potent mixture of genetic and environmental interactions facilitates the uptake of these elements into a tree's tissues. The trees in the Red Forest drew caesium and strontium up into their roots, and they then travelled to their trunks and branches. As time passed, these elements accumulated in the tree needles, which fell to the ground and became part of the forest floor litter. It's thought that caesium inhibits tree growth by impairing potassium uptake in their roots.

The trees and forest have helped to stabilise the contamination, which would otherwise have been spread far and wide by wind or water. Indeed, with the shadow of radionuclides hanging over them, these extraordinary trees stand tall, steadfast as defenders of the region. They prevent the radionuclides from spreading to ecosystems in the wider area and human settlements in, for instance, Kyiv and other areas of Europe.[6]

A view of the Chernobyl Exclusion Zone, where life is naturally regenerating. Note the Ferris wheel in the abandoned Pripyat amusement park.

I explored the mysterious world of Chernobyl's forests with Helen Whitehead, a bioacoustics specialist from Salford, Greater Manchester. Our conversation left me in awe of her passion for studying the acoustic intricacies of the natural world. Helen has a keen interest in the melodious symphonies of birds and their behaviour, and her work revolves around utilising bird acoustic recordings to monitor bird behaviour and conservation. This is similar to the soil ecoacoustics work I spoke about in Chapter 5.

During Helen's PhD, she was supervised by Mike Wood, whose work on wildlife is synonymous with the Chernobyl Exclusion Zone. Over the years, Mike has travelled to Ukraine and installed audio devices throughout the Red Forest and wider exclusion zone. He's collected a massive body of recordings of wildlife vocalisations. Many of these ended up sitting on a hard drive collecting dust somewhere at the University of Salford, and nobody had analysed them – until Helen came onto the scene. She spent countless hours scrupulously sifting through thousands of recordings. She and her Salford team wanted to know if they could assess the impacts of the nuclear accident on the bird communities (and other wildlife) in the area. In their pursuit of this, they carefully observed and meticulously collected data, aiming to unveil subtle shifts in bird populations and the wider ecosystem. During our conversation, Helen shared with me a common misconception – the notion that wildlife populations in the area continue to suffer owing to high levels of radioactivity. She said the truth of the matter is that we lack concrete evidence to support or refute this claim.

While conducting their fieldwork in the Exclusion Zone, the intrepid researchers divided the sprawling landscape into zones based on radiation gradients: low, medium and high. In each of these areas, they strategically placed acoustic recorders to capture the soundscape so they could analyse the cacophony of bird vocalisations and unlock the secrets of this mysterious ecosystem. Helen said, "People find it strange when I tell them that some areas of Chernobyl have the same radiation levels as the UK." I guess (hope) she's referring to the zone denoted as 'low'. Did the radiation affect the behaviour and long-term conservation status of bird populations in the Exclusion Zone? It seems not, at least according to the acoustic evidence. Helen and her team found that despite the common perception that the Exclusion Zone is a barren land, the diversity of bird species was actually high. They found large populations of majestic white-tailed eagles, cranes, nightjars, blue tits, blackbirds, hoopoes, cuckoos and many others, all seemingly thriving in the radioactive landscape. Helen told me that the beneficial impact on wildlife populations is largely down to the natural regeneration process. This has been allowed to happen because humans have left the area alone. Although I feel deeply sad for those people who once called the area home, it's quite a phenomenon. Despite the toxic dangers of radiation, other life has seemingly found a way to flourish.

Helen told me that people began visiting the Exclusion Zone during the COVID-19 pandemic for the thrill of being somewhere perceptually dangerous, and flocks of Instagrammers and YouTubers filmed their escapades. This was not only potentially dangerous for these attention- and thrill-seekers, but it also impacted the local wildlife. During our intriguing conversation, Helen also touched upon a sombre topic – the ongoing war in Ukraine. As we discussed the implications of this conflict, she noted that while it may deter those daring thrill-seekers, it may also usher in a whole new set of threats – involving bombs and raging fires. It's safe to say this landscape is overdue its share of solace. I asked Helen what had surprised her the most about the bioacoustics field. She said, "I think I was really surprised just how competitive the field is, but also how much information we can glean by simply leaving out tiny recording devices in trees and letting them do their thing." She also mentioned a study suggesting that radiation may cause cuckoos to produce an altered number of syllables in their vocalisations.[8] Whether this affects breeding behaviour or success is yet to be determined. I find this fascinating. I suppose the idea is that the radiation causes DNA mutations that lead to changes in the innate patterns of the bird's dialect. Here's to hoping the favourite chat-up lines of male Ukrainian cuckoos still woo the ladies and that they continue to procreate in these naturally regenerating radioactive forests.

While pondering the Red Forest, I decided I also needed to speak with someone who knew about the trees in the Chernobyl Exclusion Zone. For this, I was lucky enough to interview Norwegian forest ecologist Line Nybakken, who has worked on the pine trees in the Red Forest. When I say lucky, I mean the stars truly aligned for our meeting. I'd called on social media to see if the vast community of scientists online could put me in touch with a Chernobyl forest specialist. This didn't quite go as viral as the thrill-seeking Instagrammer posts mentioned earlier, but I received quite a few responses, one from Nybakken. This, I thought, was relatively lucky. However, the icing on the cake was that a few weeks earlier, I'd been commissioned by Jorg Sieweke to work in the Landscape Architecture department of the Norwegian University of Life Sciences (NMBU). This required me to visit Norway to run some lectures at the university. And where does Line work? Why, at NMBU of course. I couldn't believe my luck. I'd already booked my train tickets from London to Norway, and now I could 'free two birds with one key' and interview Line about the Chernobyl forests in person.

When I arrived in Norway, it was early December and the hillsides and treetops glistened with snow. Upon entering Oslo, I instantly felt I was going to like the city. It's a picturesque and vibrant metropolis located on the Oslofjord, a crystal-clear water body surrounded by verdant hills and forests. There's a pristine harbour and its architecture is a fascinating blend of old and new, with old castles and cathedrals standing alongside sleek modern buildings with bio-mimicked features (architectural designs inspired by nature). Much as in other Nordic cities, I felt an intrinsic sense of awe for nature. I can't put my finger on exactly why. Perhaps it's a combination of the rocks, the water, the hills, the trees, the geological forces and the inspiration these provide to their architects. Unfortunately, I had to leave Oslo after a day as the NMBU campus was located further south in a little town called Ås (pronounced 'Oss'). The campus was beautiful – with plenty of trees to get me in the right head space for my interview with Nybakken.

Line told me that NMBU is a Centre for Environmental Radioactivity. This research centre focuses on studying the impact of radiation on the environment and developing innovative methods for cleaning up and restoring contaminated areas. The scientists based there collaborate with national and international organisations to conduct research, provide education and training, and develop strategies for effective radiation remediation. At the time that Line and a couple of other plant people joined this group, no one in Norway had worked on plants and radioactivity, so they made use of their background in ultraviolet radiation (the kind that comes from the sun) and its effects on plants. Line specialised in understanding how such radiation causes oxidative stress to plants and was the only ecologist in the group. I asked Line how her work at NMBU related to the Chernobyl nuclear accident. She said, "Our research group has strong connections with universities in Ukraine. They have a laboratory in the Exclusion Zone, and we have co-run field campaigns." She told me that they sampled many pine trees in the area. Many of the trees were planted and some had naturally regenerated. However, "their roots go down into the dead radioactive plant material and suck up the radioactive particles", she explained. Much like Helen Whitehead's research team, Line's team took samples of pine needles from areas with low, medium and high radiation. Some analyses were done in the field and others, such as DNA analysis, were done in the lab. The research team had to transport the radioactive pine needles back to Norway, but the material was so radioactive that they could only take small amounts out of the zone.

Line told me they could see a higher level of radiation damage in the shoot tips of the pine trees from the high and medium radiation zones. She said that their needles "also produce more resin ducts as a potential protective effect". Resin ducts are tiny canals that secrete oily substances: they're often used to measure a tree's defence capacity against pathogens and other stressors. The resin acts as a warning signal. When a plant is physically damaged, it shuttles these chemicals through the ducts to protect itself.

I asked if the researchers noticed any other abnormalities in the trees that were exposed to radiation. Line said, "We also see deformities in the pine needles in the high and medium radiation zones. If you look at the trees in these zones, the regenerated ones lack apical dominance, meaning they grow much bushier than normal pines. As the young trees began to develop, an enzyme that helps to create the typical single-stem conifer shape is hindered by the radiation present in the soil. This resulted in an area with numerous dwarf pines that looked unusual and bushy. But we don't know if this is good or bad – we're unsure if it affects their health and survival."

This reminded me of Helen's work on bird populations. The radiation doesn't seem to affect the presence of wildlife, but we're still

A pine tree in an area of the Exclusion Zone with medium radiation levels – note the characteristic (normal) shape of the tree. (Image taken by Dag Anders Brede and provided by Line Nybakken.)

A pine tree in an area of the Exclusion Zone with high radiation levels – note the deformed 'bushy' shape of the tree. (Image taken by Dag Anders Brede and provided by Line Nybakken.)

not sure whether it harms them in any meaningful way. Essentially, we lack qualitative rather than quantitative data.

I asked Line what had surprised her the most about the Chernobyl work. She replied, "One of the things I found most surprising is that even though the radiation levels in the high and medium zone are far lower than those in the immediate aftermath of the initial accident, we're still seeing major reactions in plants 30 years later. It suggests that chronically low radiation levels still change the trees' DNA in all kinds of weird ways. In particular, small conifer shoots are highly affected by very low doses. However, at the same time, other wildlife seems to be thriving here." Line also told me of a few more light-hearted but still stressful things that happened during her field campaigns at Chernobyl. She said, "You never know what will happen when you go there. In 2018, we couldn't access the area because of fires. So we became tourists for a few days and then had to work like demons to get the fieldwork done quickly when the fires had dissipated."

Line's point about fires got me thinking. How might building fires affect the radiation risk in the zone? And how might forest fires affect this?

Remember how I said that the forests had soaked up the radioactive particles? This is where climate change and increased risks of forest fires pose a considerable problem.[8] I asked Line how climate change might affect this radioactive particle absorption issue. She said, "The remarkable capacity of the forests to endure and limit radiation is jeopardised by increasingly severe wildfires, which are exacerbated by the effects of climate change. The trees, which were once, in addition to being intrinsically valuable, a helpful presence for human health, now become a threat when they catch fire. The burning of the trees releases the radioactive substances they have stored, causing them to be dispersed as aerosols that humans and other animals can inhale. Moreover, the contamination now comes from the trees that blanket an area of over 160,000 ha surrounding the Chernobyl nuclear power plant instead of emanating from a single source as it did in 1986. In other words, the risk of radioactive exposure is now much more widespread and difficult to contain. While the planted and naturally regenerated forests have been our friends in containing radiation, they may become a threat when they burn. Therefore, their survival is more critical than ever."

So, while natural regeneration may be taking place, and from an external viewpoint the forests may appear to be flourishing and expanding, there lurks a concealed yet profound peril beneath the surface. For 30 years, trees have been cleaning up our highly toxic mess. But this mess (radioactive particles) has a very long half-life. This is the time it takes for half of the radioactive atoms in a sample to undergo radioactive decay (i.e., lose their energy). Different radioactive isotopes have different half-lives, ranging from fractions of a second to billions of years. For example, the half-life of iodine-131, a radioactive isotope used in medical treatments, is approximately eight days. In contrast, the half-life of uranium-238, a common radioactive isotope found in nature, is about 4.5 billion years.

In addition to uranium, these are some of the radioactive isotopes in the Chernobyl Exclusion Zone and their half-lives:

- Caesium-137: the most abundant and long-lived radioactive isotope in the zone. It has a half-life of about 30 years and was responsible for a significant portion of the contamination.
- Strontium-90: another long-lived isotope with a half-life of about 29 years. It's like calcium and can accumulate in bones (as in trees), posing a health risk to humans and animals.

- Plutonium-239: a highly toxic and long-lived isotope with a half-life of about 24,000 years. It's a by-product of nuclear reactions and can cause cancer and other health problems.
- Americium-241: a radioactive isotope used in smoke detectors that can also be found in nuclear waste. It has a half-life of about 432 years and can pose a danger if ingested or inhaled.
- Iodine-131: a short-lived isotope with a half-life of about eight days. It can accumulate in the thyroid gland and increase the risk of thyroid cancer.

This is some 'toxic mess' that the trees endure, and it takes a very long time for the dangerous stuff to dissipate. But it also comes with a potent and sobering message about the risks of climate change, driven in part by our insatiable desire for commodities.

I've discovered that the forests have seemingly thrived in the wake of a massive nuclear accident. Wildlife populations have bounced back, birds are still singing songs and trees are growing but look deformed. But it's crucial to critique the "all's fine and rosy" narrative; after all, we can't ask the trees how they feel, at least not linguistically. As such, is there any other evidence that wildlife and trees suffer rather than thrive?

I went once again on a literature treasure hunt. The first paper I found referred to cataracts.[9] This medical condition affects the eyes and causes clouding of the natural lens, which is usually clear. As the lens becomes cloudy, it can cause blurry or dim vision, sensitivity to light, difficulty seeing at night and other visual problems. Cataracts typically develop slowly over time and may initially have a minor impact on sight, but they can eventually lead to significant visual impairment. The study found that voles (small rodents and relatives of lemmings and hamsters) that live in and around the Red Forest had a much higher frequency of cataracts than natural populations. In the female voles, this correlated with accumulated radiation dosage. Moreover, female voles with more severe cataracts tended to have fewer offspring. The study suggests that high levels of oxidative stress may cause cataracts owing to ionising background radiation. Therefore, the increased background radiation in Chernobyl could have direct effects on the fertility of local mammals in and around the Red Forest and indirect effects through increased cataract development.

Other research suggests that bumblebee reproduction and colony growth are delayed in the most contaminated areas of the Exclusion Zone.[10] This could negatively affect pollination and overall ecosystem function. The researchers' lab-based study also demonstrated that exposure to radiation levels similar to those found in the Exclusion Zone decreased the number of queens produced by bumblebee colonies. This reduction ranged from 30% to 45% compared with unexposed colonies. Queens are crucial for the survival of bumblebee populations as they are the only ones that reproduce: worker bees only help with food collection and maintenance of the colony. After hibernating over winter, the queen starts a new colony and produces workers until late summer, when she produces either male bees or new queens. As the new queens are larger and more resource-intensive to produce, more queens produced means more bumblebee nests will be present the following year. A decline in queen production is a common response to stress, as seen in insecticide exposure, and can significantly impact the survival and growth of bumblebee populations. Therefore, radiation, particularly in the Red Forest area where it's higher, could be harming queen bees and the ecosystems that depend on them. Another study showed that some birds in the Exclusion Zone have smaller brains than those outside.[11] The researchers captured over 500 birds using large mist nets in the forest. They inferred brain size by measuring their head volume, which is apparently a robust proxy. They found that low-dose radiation can significantly affect normal brain development, as reflected by brain size. Therefore, the Chernobyl radiation may affect the cognitive ability of the wildlife. However, it's important to point out here that the authors of the last study have been criticised for underestimating the radiation doses that the organisms they studied had been exposed to.[12] So perhaps we should take their findings with a pinch of salt.

From Helen's bioacoustics work, it seems that bird song is not significantly affected by the radiation, nor is the diversity of bird species present in the Exclusion Zone. But studies showing mutations that might lead to physiological deformities and issues with reproduction suggest we need to consider the potential hidden health impacts of the nuclear disaster. It's a consideration that falls within the realm of environmental ethics.

I'm now scouring my memory to reflect on the days of learning about the likes of Immanuel Kant, Aldo Leopold and Arne Naess in my undergraduate ethics classes. From a biocentric perspective, all living beings, including non-human animals, have inherent value and should be protected regardless of their utility to humans. My old ethics lecturers might say that preventing wildlife from entering their natural habitats in the Exclusion Zone violates their intrinsic rights to exist and thrive. On the other hand, allowing wildlife to enter areas (particularly the Red Forest area) that we've contaminated with dangerous radiative materials might be considered ethically wrong.

The effects of radiation on wildlife in the Chernobyl Exclusion Zone are still being studied, but there's evidence that some animals have been negatively impacted. On the other hand, some mammal populations, such as wolves and wild boars, have increased in the area owing to the lack of human interference. Still, it's unclear how the radiation has affected their health and longevity. Additionally, some species of insects and plants have shown signs of radiation-induced mutations. However, some species seem to be thriving in the absence of unsustainable human activity, such as the endangered Przewalski's horse, which has successfully been reintroduced to the area.

There's a divide among scientists regarding how radiation has impacted the wildlife residing in contaminated areas. Perhaps a more pertinent question for this chapter is: Given the radioactivity, the mutations and the deformities, is this *natural* regeneration?

CHAPTER 8

Agroforestry

"Life is a succession of lessons that must be lived to be understood. All is a riddle, and the key to a riddle is another riddle."
—Ralph Waldo Emerson

In the 1970s, a young boy named Klaus grew up in a suburban area of Germany. The design of the German suburbs was based on a model of resilience. Every half-acre was planted with fruit trees and potatoes, and the streets were peppered with chicken coops and goat stables – a far cry from today's Western European urban environments. However, this suburban model had a dark history: it was part of a Nazi masterplan. A few decades before, the Nazis had prepared for a prolonged war and wanted to create a system of self-sufficiency for the country's urban communities. For young Klaus, the bright side of this masterplan was being schooled in the techniques of permaculture. Despite its tenebrous foundations, he lived in a flourishing permaculture suburb.

Permaculture can be viewed as a way of life and a land management approach. Indeed, its practitioners adopt arrangements observed in natural ecosystems to establish an ecologically sound way of living. Its principles extend to diverse aspects of life, including energy systems, architecture, waste management, water conservation and community building. However, cooperating with nature and caring for the planet and its people are central values.

After Klaus finished high school, he travelled to South America, and while travelling through Brazil he landed an internship with the Swiss farmer and researcher Ernst Götsch. Ernst had moved to Brazil from Switzerland in the 1980s, establishing a farm in Southern Bahia. The land he purchased for his farm had been "hopelessly degraded" by traditional intensive farming practices in previous years. Ernst started to develop a new farming technique that incorporated deep ecological principles, and after a few years, the land was surprisingly restored with high levels of biodiversity and productivity, a win–win situation for humans and non-humans alike. Astonishingly, the biodiversity levels were comparable to virgin forest in the surrounding Amazon. Ernst persisted in honing his ecological farming approach, tirelessly refining and evolving it. Today, his pioneering method has become widely known as syntropic agroforestry.

When Klaus joined Ernst in Brazil, his permaculture upbringing chimed strongly with these syntropic agroforestry principles. Despite Klaus's permaculture skills, he jokes, "I was mostly hired to look after Ernst's son, who only had four sisters to keep him company." After a while, Klaus refined his machete skills to clear scrub, as burning was not considered an appropriate method to control the vegetation. He worked with Ernst to plant bananas and cocoa trees in a permaculture style in the degraded soils, but this didn't work. In a practical sense, permaculture is a design framework with limitations. While encompassing various sustainable practices, it may not always directly facilitate the replication and acceleration of natural processes such as natural succession (the process of change in the structure of an ecological community over time) and stratification in degraded habitats.

To do this, a deeper knowledge of ecological principles is required, bestowing each plant with the ideal conditions for its development. This is where syntropic agroforestry comes in. This fosters a deep understanding of where to place and cultivate each plant optimally in space (strata) and in time (succession). It also requires an acute understanding of the soil – the foundations of the ecosystem. Let's take a quick look at the term syntropic or syntropy. It's the opposite of perhaps a more familiar term, entropy, which is often mentioned in relation to the laws of thermodynamics. Entropy classically refers to the disorder of a given system and is linked with energy dissipation; it's also thought of as a lack of order or predictability. In contemplating the natural tendency of energy to dissipate and simplify, it's essential to acknowledge the presence of a complementary tendency or force that counterbalances this trend. This is syntropy, which manifests as ordered structures, complexity and differentiation, much like the natural order or the evolution of life. Therefore, life in our complex ecosystems is a manifestation of the syntropic force.

For many decades, the notion of a complementary force to entropy has intrigued mathematicians and scientists. This concept underscores the methods in syntropic agroforestry. It's like speaking the language of nature to simultaneously enhance biodiversity and crop productivity. However, the dominant agricultural model is to suppress the natural order of ecosystems to produce monocultures of annual plants, such as rice, wheat and barley. By doing this, we work against nature and often degrade the precious soils that sustain healthy ecosystems. Syntropy is the reverse of this degradation. Not only does it embrace natural succession, but it also actively promotes it.

The Lotz-Keegan family

I interviewed Klaus via video call in 2022. When we spoke, he was sitting outside on his family's syntropic farm in Aotearoa (New Zealand) in the evening; I could hear the buzzing of the insects in the background. Klaus moved to Aotearoa a couple of decades ago, and the north island is perfect for growing a wide range of crops. The entire family work the farm, including Klaus, his wife Vanessa Keegan, his daughter Frida and his son Josh. Each has a unique skillset that helps the farm run like clockwork. For instance, they're all farmers, but Frida is also a tutor and landscape designer, Vanessa is a nutritionist and horticulturalist, and Josh is a carpenter and builder.

I wanted to know more about syntropic agroforestry and how it could help restore forest ecosystems while simultaneously empowering human livelihoods. To this end, Klaus told me about three core principles of a successful syntropic agroforestry system.

Principle one: strata

This means capturing as much sunlight as possible before it reaches the ground. To do this, it's essential to understand the different canopy layers that plant communities occupy. It's not so much the plants' height that determines these canopy layers but the degree of sunlight they require at maturity. The top layer is the emergent canopy, followed by the high, medium and low layers. There's a classic example of this for annual crops, also known as the three sisters, in which a pumpkin plant, a bean plant and a corn plant are grown together. The pumpkin plant sprawls out and occupies the mid- to lower strata, then the corn plant shoots upwards, standing tall and occupying the emergent layer, and the bean plant grows up the corn plant, occupying the mid- to higher layer. In most plant communities, you'll find that they occupy their own niches, thus preventing competition. The degree of frost tolerance is also important when considering strata: the emergent canopy is often the most frost-tolerant and offers protection to the lower levels of vegetation.

Principle two: lifecycle

Plants can either be annual (growing, living and dying over the course of a year), biennial (completing their lifecycle over two years), short-lived perennial (3–10 year lifecycle), long-lived perennial (10–20 year lifecycle) or primary (20–40+ year lifecycle). Understanding the lifecycle of plants grown in a syntropic agroforestry system is vital to creating functional diversity over time and space. For example, in the first year, you may plant an avocado tree in among a tamarillo tree, an ingo bean tree and a cherimoya tree, while annual plants (such as the corn, bean and pumpkin) will grow in the low strata layer. If we fast-forward by 20 years, you'll find the mature and productive ingo bean canopy in the emergent layer and the mature fruiting avocado canopy in the high layer, nestled beneath the ingo bean tree, and the fruiting cherimoya tree nestled beneath these in the medium-to-low strata layer.

By incorporating careful strata and lifecycle considerations, this planting arrangement fosters a flourishing and diverse system that thrives from its inception and extends well beyond two decades.

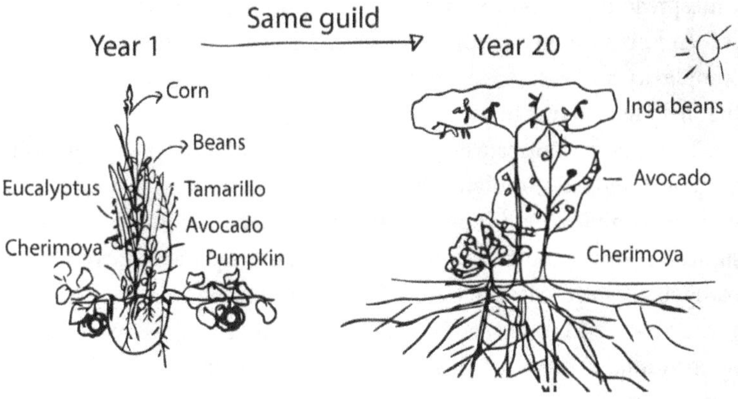

Year 1 and Year 20 of a subtropical syntropic agroforestry system (adapted from PermaDynamics).

Principle three: succession

This can be broken down into four phases: the establishing phase, the accumulation phase, the transitionary phase and the abundance phase. The vegetation in each phase differs and is influenced by time and soil fertility. For instance, you may see lichens and mosses in the establishing phase as they erode bare rock to generate soil. In the accumulation phase, there are plants such as thistles, gorse, tea tree and conifers, which don't produce fruit but play an important role in preparing the soil and accumulating nutrients. In the transitionary phase, the plants become lusher as fertility increases, and in the abundance phase, the vegetation is lush, fruits are abundant and the soil is highly fertile.

A deep understanding of when and how plants need to be pruned is also essential for syntropic farming. This includes so-called weeds. Klaus said they have a saying on their farm, "Weeds to redundance leads to abundance." Essentially, unlike most farming practices, they keep the weeds until they are redundant. These are primarily opportunists that need an abundance of sunlight; they help accumulate nutrients in the soil and they attract pollinators. The key is, Klaus told me, to prune them to get another growth, build the nutrients in the soil and stimulate them to thrive, "whatever gets the biomass pumping". Eventually, there's too much shade for them as the other vegetation grows and the weeds are no longer present – they're redundant. Klaus called them "helper plants", a refreshing perspective given the widespread demonisation of weeds globally. They help to prepare the soil and help the mycorrhiza to thrive. This enriches the system with complexity and resilience. Pruning also stimulates growth hormones that help new plants to grow.

Pruning a plant can be compared to a workout for the body. In humans, exercise stimulates the production of growth hormones that help build muscle and increase strength. When a plant is pruned, it signals to the plant that it needs to produce more growth hormones to compensate for the loss of tissue. This results in a burst of new growth, as the plant sends out new shoots and branches to fill the space left by the pruning. Over time, this new growth can help the plant become stronger and more resilient as it develops a more robust and healthy structure. Much as the body needs rest and recovery time after a workout, plants also need time to recover after pruning, so they can fully utilise the growth hormones and produce new growth.

Klaus told me that most societies have become attached or "even addicted" to farming annual plants. We work against processes of natural succession, which destroy ecosystems and our social and spiritual connection with the natural world. He firmly believes our attachment to monocultures of annual plants (such as wheat and rice) has led to society's downfall, the root of many global problems. Succession is natural, it's a sculpture that forms all around us, but we see it as our enemy most of the time. He said that syntropy provides a language to understand how plant communities restore the soil. Diversity is key. Diversity in plant species brings a diversity of crops and a diversity of wildlife.

Speaking of wildlife, several years ago, Klaus kept getting woken up by a noise in the leaf litter outside. One night, he ventured out into the forest with a green hunting light to find what was making the noise. He discovered it was a kiwi – New Zealand's national bird. All kiwi species are endangered due to three main threats: non-native predators, habitat loss, and fragmentation of populations. Klaus says the kiwis are now thriving in the syntropic agroforestry system, or in his own words, "We can go outside to take a piss and stumble over kiwis". They seem to love the softer soil, which is packed full of worms. There's also a rare native duck that visits to feed on fruits. Birds and insects flock to the forest because of its diversity of vegetation and niches across space and time. That the farm is a biodiversity haven is perhaps a strange thing in today's climate, but life abounds here – a vibrant scene. Klaus paints a picture in my mind of bees humming, butterflies dancing amid the swaying flowers and worms crawling in the humus, all working in tandem to create a thriving ecosystem.

Klaus told me that, in terms of economics, the lucrative 'cash crops' are the ones that need no input. No spray or labour: simply leave them to do their thing. Cherimoya is one such crop. This subtropical fruit from the Andes is an ancient plant with a weird rudimentary flower that can heat itself during the night to attract pollinators. This is because it didn't evolve attractive colours in its flower, the kind that usually draw bees and flies in to gather pollen. However, pollination is the main problem for growers of cherimoya; many resort to hand pollination, which is time and labour intensive. Klaus tells me this is not an issue in the syntropic farm. He owes this to the high diversity of plants and many native ones that attract swathes of insect pollinators. Moreover, pesticides aren't used, which is obviously favourable to insects. "Although the individual plants may yield 10% less than those

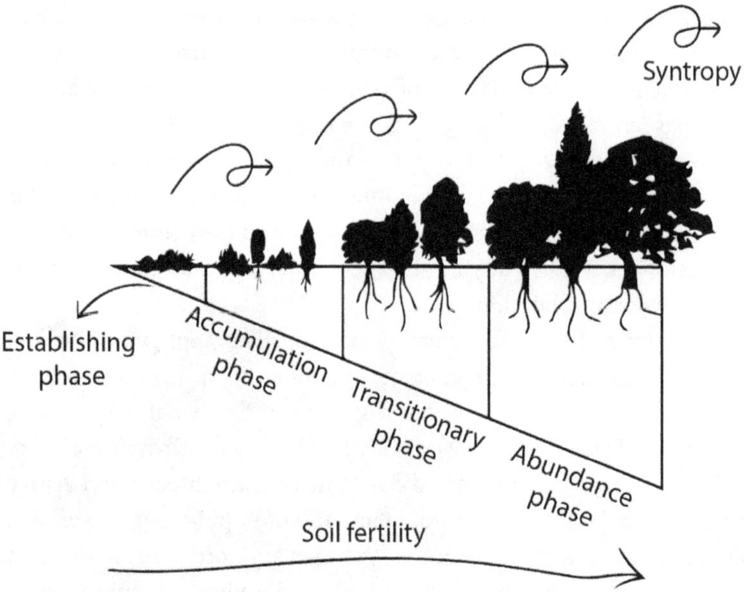

Succession in a syntropic agroforestry system.

grown in monoculture plantations, when you calculate the yield of all the different plants grown in the forest, we probably double the yield of produce compared to a monoculture covering a similar area of land."

A monoculture, by definition, lacks the diversity and resilience of a mixed crop system. Only one type of crop is grown, which can lead to a lack of genetic diversity and an increased susceptibility to diseases and other stressors. In the same way that a choir needs a variety of voices and harmonies to create a beautiful euphonious blend, a mixed-crop system needs a variety of plants to create a healthy and sustainable ecosystem. By planting a diverse array of crops, farmers can promote beneficial insects and microbes, build soil health and reduce the need for harsh pesticides and fertilisers.

It's not just plants that thrive under syntropic farming – fungi do too. Klaus's family harvests fungi for medicine and food. One tree is close to Klaus's heart because it's like a tree he loved when he worked and lived in Bolivia: it's in the mahogany family and is called the kohekohe tree in the Māori language. It grows up to 15 m high and has oval green leaves, similar to the European beech. However, we're interested in fungi here, and the tree is host to a particular mushroom called a *Ganoderma*. This has been employed in Chinese traditional medicine for thousands of years to treat a range of maladies and boost the immune system. The family also deliberately cultivate another fungus in the syntropic forest: the famous shiitake mushroom. These are best grown on hardwood tree species, and whenever the family works in the food forest, they identify the most suitable branches. They inoculate the branches with shiitake mushroom spawn, and once this is embedded into the logs, it spreads out and colonises them. The logs are then laid out on the forest floor, and eventually they produce fresh and beautiful mushrooms. Depending on the diameter of the log, the mini-mushroom factory will last for between two and nine years. Shiitake mushrooms are very popular, and Klaus's family grow them uniquely and sustainably – usually, they're grown in controlled and sterilised grow rooms. Once the lifetime of the mushroom log is over, it's returned to the forest floor, where it decomposes and replenishes the carbon in the soil. I asked Klaus whether 'everyday people' could apply syntropic agroforestry methods to their gardens. He said, "Sure they can. We have one plot that is only 6 metres by 8 metres. Syntropy doesn't need a 1,000-year goal, although it could be transferable to long-term restoration contexts; it could simply be a five or ten-year project where something new is started after." As long as the three core principles are considered (strata, lifecycle and succession, with a healthy dose of pruning in the right way at the right time), syntropy can be used to create a flourishing garden that enhances biodiversity. It also has the benefit of producing delicious crops and improving soil health. This requires a degree of research – for example, identifying the right plant species and how to tend to them – but if you're looking for a healthy challenge and seek to strengthen your connection with the natural world, syntropic gardening might be for you. I know my mind is now fizzing with ideas for our garden at home. Loyal to the permaculture ethos, Klaus's family lives an ecologically sound life. Even the structures on the farm are made from the local land, and when their lifetime is up, they're returned to the land to feed into the natural cycles. For instance, Klaus's son Josh built a drying hut to remove the water content in fruits. It was constructed out of clay, straw and wood. One wall took him 40 minutes to make, and if it needs to be removed or if it degrades, he told me "You can simply put it in the compost as opposed to something that needs to go to a landfill." Inside the drying hut, there are several shelves for the trays that hold the fruit and a wood stove at the back to heat the room. There's a beneficial by-product from this drying process. Known as biochar, it's the carbon that remains from wood when it's burnt in a controlled process called pyrolysis. When it's mixed with soil, biochar aids in retaining water, minerals and trace elements that are usually flushed out of the soil. Essentially, it enhances the soil's sponge-like abilities, while also providing a habitat for various beneficial soil microbes that promote soil and plant health. The Lotz-Keegan family also produce biochar when they heat their hot tub by burning the non-native gorse plants that grow on their land.

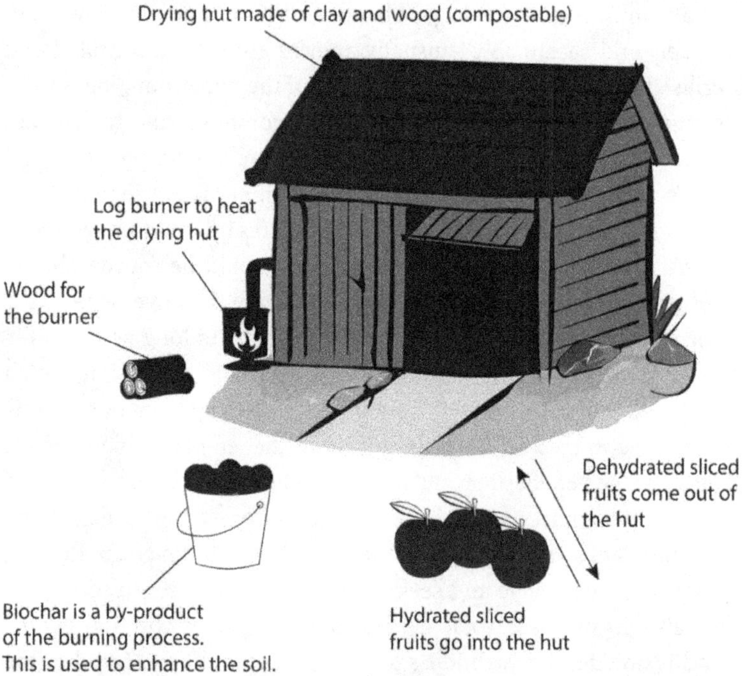

Drying hut on the syntropic agroforestry farm.

I asked Klaus if there was anything else he'd like to mention. He said, "Only that we need to look at syntropy and food forests as being more dynamic than our current dominant food growing models. Nature is not static. Therefore, we shouldn't make a static plan. It's like a child; we can't keep it as a two-year-old forever just because it's cute or cheap. It's the same with a food forest. Nature is constantly evolving, so it's wrong to keep it static. We need the dynamics to shake things up. And these shake-ups stimulate life. Humans have become too good at being disturbers. We've overshot our design goal. Annual farming is fucking up the planet. We need to align with regeneration; it's incredibly powerful. Most agriculture is about working against succession. Syntropy brings back energy and diversity!" I wondered whether syntropic agroforestry methods could be transferred to a traditional ecosystem restoration context. For instance, could we think more about strata, lifecycle, succession, use helper plants and apply the right pruning to stimulate growth hormones? Klaus said, "Of course. One of the key benefits of the syntropic method is to bring biomass and diversity back to degraded land." Klaus is looking into riparian (riverbank) restoration. Rather than keeping field margins between cattle and the river minimal, why not create a marginal habitat to bring nutrients to the animals? Having certain tree crops in the zone that are beneficial for the land, the river and the animals could be hugely restorative.[1]

Other forms of agroforestry

While the focus of this chapter has been on syntropic agroforestry owing to its innovative approach and deep ecological principles, other types of agroforestry exist. At a basic level, agroforestry is the establishment of trees with agricultural crops and/or animals on the same land. Agroforestry systems are designed to provide crop products while simultaneously protecting, conserving and diversifying environmental 'resources'. The Agroforestry Research Trust says that agroforestry practices differ from traditional forestry and agriculture by focusing on "the interactions amongst components rather than just on the individual components themselves".[2] Research has shown that agroforestry systems can have tremendous benefits over traditional agricultural and forestry systems. They can control run-off, reduce soil erosion and maintain soil biological activities, vital for fertility.[3] This requires a considerable proportion of tree cover in the landscape (20% or more). Restoring trees in the agricultural landscape can help create a closed nutrient cycling system, where synthetic chemical fertiliser input is unnecessary. These systems can lead to a reduction in 'pests' and diseases. It's context-dependent, but agroforestry systems can also enhance biodiversity.[4] According to the Agroforestry Research Trust, agroforestry systems may only use 5% of the agricultural land area but account for more than half of the biodiversity in the farming landscape.[5] The diversity of species brings a variety of functions to the ecosystem.

A complex and diverse ecosystem is more likely to be resilient in the face of significant environmental challenges such as climate change. And speaking of climate change, the additional trees in the landscape contribute to climate change mitigation by sequestering carbon and restoring ecosystem integrity. According to the Climate Institute, agroforestry systems possess equivalent or sometimes higher organic soil carbon than natural woodland plots.[6]

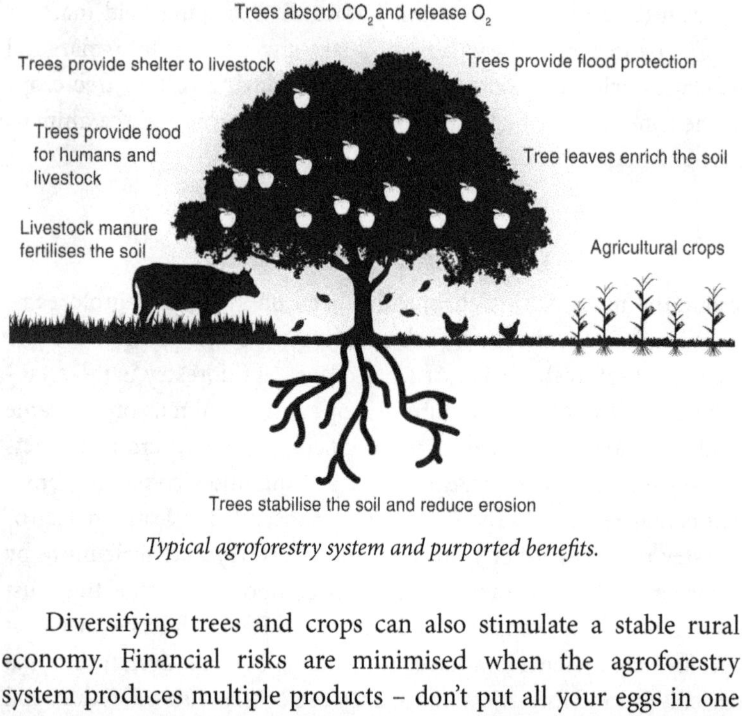

Typical agroforestry system and purported benefits.

Diversifying trees and crops can also stimulate a stable rural economy. Financial risks are minimised when the agroforestry system produces multiple products – don't put all your eggs in one basket, as the saying goes. The shelter provided by trees improves yields of nearby crops and livestock and can reduce the heat stress on livestock during the warmer months of the year. And in terms of forest restoration, agroforestry systems have immense potential to contribute to treescapes. They can provide corridors for wildlife and connect different forest ecosystems, thereby reducing forest fragmentation in the landscape. As you can see, there are many benefits of an agroforestry system, but are there any disadvantages?

First, what works on one farm, may not work on another. Because the focus is on interactions and ecological principles, it can be quite a complex process to ensure the components of a system harmonise. It can also be labour-intensive and requires adequate knowledge, planning and maintenance. Perhaps the biggest perceived disadvantage to agroforestry systems is time.

Agroforestry isn't a 'quick fix' because trees take several years to grow and mature. If the trees are being used for crops themselves, for example, to produce nuts, it may be ten years before the tree reaches its full production level. However, a single nut tree can produce nuts for several generations – over 100 years.[7] Therefore, a long-term vision is sometimes needed, an embracing of delayed gratification. However, as shown in the syntropic examples, you can grow annual plants instantly. Moreover, a deep understanding of lifecycles, strata and succession allows farmers to harvest various crops each year. As Klaus mentioned, we became particularly addicted to annual monoculture plantations because they were viewed as the most productive. Therefore, policies and incentives most often favour these low-diversity systems.

As the environmental journalist George Monbiot suggests, the ubiquitous global agricultural system (what he calls the global standard farm), found on nearly every continent, follows a standard model of cultivating common crops and livestock using standardised genetics, fertilisers and pesticides.[8] However, this uniformity in agronomic practices is leading to the degradation of soils worldwide, as different areas face similar challenges ranging from drought to disease. Moreover, the global standard farm is also the single biggest source of greenhouse gas emissions – so the food system risks killing us as much as nourishing us.

Unfortunately, policies that support agroforestry systems are not comprehensive or widespread. The first country to adopt an agroforestry policy was India in 2014, with its National Agroforestry Policy.[9] This is a "comprehensive policy framework designed to improve livelihoods by maximising agricultural productivity for mitigating climate change".[10] Agroforestry is growing at an unprecedented pace globally, so it's unlikely to be too long before policymakers catch up (fingers crossed!).

It's also important to select the right trees to complement other crops (and not compete with them), not cause diseases or produce plant-inhibiting chemicals. The latter is known as allelopathy. Some trees produce chemicals that can inhibit the growth of nearby plants.

For example, some Eucalyptus trees can suppress vegetation up to 10 m away by shuttling allelopathic chemicals into the ground.[11] And these were once a widespread favourite in agroforestry systems.

The land-sharing problem

I've spent countless hours walking around agricultural fields in the UK doing vegetation surveys and bat transects. If you did the same, you'd probably notice that most land is covered in a monoculture crop, such as wheat or barley. But you may also notice a narrow margin of mixed grasses and wildflowers or even a scattering of trees in the corner of a field. These strips of natural habitat in the sea of annuals are left by design, not by accident. Regions and countries such as the EU, the UK, Japan and Mexico have directed resources towards environmentally sustainable land-use practices in response to the biodiversity crisis.

The Common Agricultural Policy,[12] established in the 1960s, has been the EU's costliest policy. Over one-fifth of the budget, amounting to £60 billion annually, is allocated for 'environmental improvement'.[13] A significant portion of these funds is channelled into initiatives that promote land-sharing schemes. These aim to strike a balance between agricultural activities and environmental conservation by implementing practices that benefit both objectives. But what does this land-sharing typically look like? It often manifests as those narrow meadow strips and tiny woodland copses in the corner of a wheat field, but can also include the adoption of more diverse cropping systems – although I've personally not seen any syntropic-inspired systems in the UK.

Academics Ian Bateman and Andrew Balmford recently spoke about the pitfalls of such land-sharing schemes.[13] They suggested that from a conservation perspective, land-sharing often increases relatively common species, such as skylarks and field poppies, but does little for the threatened species that need vast expanses of old-growth forest. Moreover, Bateman and Balmford said that when land is taken out of agriculture without reducing food demand or increasing crop yields elsewhere, it leads to a greater reliance on food imports from other countries. Unfortunately, this can have detrimental impacts on biodiversity and habitats in distant locations, often going unnoticed or ignored owing to the 'out of sight, out of mind' phenomenon. They pointed to the statistic that in the 25 years up to 2014, the EU's crop imports generated more than 11 million ha of habitat destruction abroad. This is an area larger than Cuba, and is often to be found in some of the world's biodiversity hotspots.

The rewilding problem

By taking large, contiguous areas out of farming, rewilding initiatives can provide a lifeline for wildlife on the brink of extinction. Take, for instance, the story of Ken Hill, Norfolk. In an effort to rewild 400 ha of this beautiful landscape, conservationists have succeeded in creating a safe haven for beavers. These industrious creatures, once considered extinct in the UK since the sixteenth century, now roam freely and shape their environment in ways that benefit numerous other species. This is just one example of how rewilding can breathe new life into local biodiversity. Undoubtedly, rewilding initiatives hold value for national conservation efforts. They offer a glimmer of hope in preserving and restoring delicate ecosystems. However, there's a catch that often goes overlooked – and it's the same issue that blights land-sharing. When we consider the benefits of rewilding, we seldom account for the potential damage inflicted on biodiversity in faraway places.

Here's the crux of the matter, as pointed out by Bateman and Balmford: If we remove large swaths of land from productive agriculture without simultaneously reducing food demand, increasing crop yields elsewhere or changing our dietary habits, we inadvertently fuel the demand for food imports. And as the demand for imported food rises, it takes a toll on biodiversity in other regions; again, out of sight and out of mind. It's a complex conundrum. While rewilding can provide localised benefits and serve as a refuge for vulnerable species, we must balance this against the unintended consequences. Suppose we're to truly protect biodiversity and ensure the long-term sustainability of our planet. In that case, we need to find ways to reconcile our land-use practices with global food demands. This may involve embracing sustainable agricultural practices, promoting responsible consumption and striving for increased yields in the areas that remain under cultivation.

The organic problem

Another popular approach that Bateman and Balmford have posited will lead to greater demand for food imports and hence affect biodiversity around the world is a large-scale switch to organic farming.

Organic farming (a system that uses organic fertilisers, crop rotation and companion planting) has been around for millennia but fell out of favour as it was replaced by massive and efficient monoculture agriculture. It has, however, regained significant attention in recent years. Both the EU and Japan have made bold commitments to convert a considerable portion of their farmland to organic production. The EU aims to achieve this transformation by 2030. Japan has also set an ambitious target of reaching 25% organic farmland by 2050.

The allure of organic farming lies in its promise to eliminate synthetic fertilisers and modern pesticides from agricultural practices. This shift is expected to benefit certain species that suffer from the harmful effects of these chemical inputs. By removing such potentially pernicious substances, organic farming offers a more nature-friendly alternative. However, it's important to acknowledge the limitations of organic farming when it comes to supporting species that require contiguous natural habitats. While the absence of manufactured fertilisers and pesticides can be advantageous for some farmland species, those reliant on undisturbed and interconnected habitats may not reap the same benefits. The fragmented nature of agricultural landscapes, even under organic management, can impede the survival and movement of certain species, limiting the positive impact of this farming approach.

A further critical consideration, Bateman and Balmford maintained, is the significant reduction in yields associated with organic production. Sri Lanka's food crisis in the 2020's serves as a cautionary tale. The high cost and scarcity of food have become major challenges for a large portion of the population. These issues are closely tied to an agricultural policy implemented by the Sri Lankan government that aims to swiftly phase out chemical fertilisers in order to prioritise or demonstrate environmental sustainability.

The government attempted to convert the entire country to organic farming. And although now abandoned, this was believed to have played a significant role in the crisis. They wanted to ban chemical fertilisers to be (or appear) environmentally sustainable, but this led to a massive decrease in food production. With this decrease came a hike in food prices, which meant people struggled to meet their daily food needs. This indicates that a complete shift to organic farming without careful planning and consideration of the potential consequences can have severe repercussions on food availability and security. Indeed, estimates indicate that embracing organic farming on a large scale across England and Wales could result in a staggering 40% reduction in food output.[14] This statistic highlights the trade-off between the environmental benefits of organic farming and the challenge of meeting the increasing demand for food in a world grappling with population growth and food security concerns.

The ongoing debate surrounding organic farming underscores the complexities inherent in our agricultural systems. While organic practices can offer certain advantages in terms of reducing chemical inputs and benefiting specific farmland species, they may fall short in preserving contiguous habitats and maintaining high food productivity levels. I also wonder if we need more trials with the syntropic agroforestry approach – which, as Klaus maintains, can increase crop yield and biodiversity. This would probably fall under the 'organic' farming umbrella but with potentially more benefits for trees, wildlife and local economies.

Is 'land-sparing' the solution?

The two professors think there's an alternative approach that holds great promise for supporting both local and global biodiversity. Their proposed solution is 'land-sparing'. Unlike rewilding and organic farming, land-sparing focuses on consolidating habitat patches into larger, contiguous blocks while implementing lower-impact methods to increase yields in the remaining cultivated areas. This combined strategy can have significant positive effects on habitat protection without displacing agricultural production to other countries.

The concept behind land-sparing is straightforward yet potentially powerful. By grouping habitat patches together, we create larger and more connected areas that can better support wildlife populations. This consolidation allows species to move freely, find mates and access resources, leading to healthier and more resilient ecosystems. At the same time, it ensures that agricultural production continues in the remaining farmland (which could be syntropic), where innovative techniques can be employed to boost yields while minimising environmental impacts. The beauty of this approach lies in its ability to harmonise the needs of both nature and agriculture. By sparing larger land areas for conservation, havens are created where biodiversity can thrive. Simultaneously, by optimising agricultural practices in the remaining cultivated regions, we can increase food production without encroaching upon additional natural habitats or relying on food imports from environmentally vulnerable regions.

Implementing land-sparing requires a careful and integrated approach that combines scientific expertise, land-use planning and supportive policies. It necessitates collaboration between conservationists, farmers, policymakers and other stakeholders to ensure successful implementation. Bateman and Balmford have noted the numerous field studies conducted in various countries that have shed light on the impact of different agricultural approaches on wildlife populations: India, Ghana, Uganda, Kazakhstan, Mexico, Colombia, Brazil, Uruguay, Poland and the UK have all been part of this extensive research effort. The findings of these studies consistently demonstrate that when considering the same overall food production, a combination of high-yield farming practices and land-sparing leads to larger populations of most wild species compared with land-sharing approaches. This is a significant revelation with implications for biodiversity conservation.

These results have been observed across a diverse range of species, including birds, plants and insects, totalling more than 2,000 different species. The benefits of high-yield farming and land-sparing are particularly prominent for species with limited geographical ranges, which are especially vulnerable to habitat fragmentation and loss. The evidence suggests that land-sparing, which involves consolidating habitat patches and optimising agricultural practices, creates more favourable conditions for wildlife populations to thrive.[15] Providing larger contiguous areas of protected habitat allows species to maintain their ecological roles and supports their population growth.

These findings challenge the notion that sharing land between agriculture and conservation is the most effective approach for safeguarding biodiversity. Instead, they highlight the advantages of adopting high-yield farming practices alongside land-sparing, which strikes a balance between food production and habitat preservation. By prioritising the implementation of these approaches, we can create landscapes that sustain both human livelihoods and robust wildlife populations. It's through a careful and integrated combination of strategies that we can achieve a coexistence between agriculture and nature, ensuring a healthier and more resilient planet for generations to come. It's crucial to acknowledge that increases in agricultural yields do not automatically guarantee that more land is available for conservation purposes. Critics of the land-sparing approach rightly point out the potential challenges that can arise, despite yield gains. One of the concerns raised is the possibility of so-called rebound effects. In some cases, when agricultural productivity improves, the demand for agricultural products may increase. This increased demand can result in less land being taken out of farming than initially anticipated. For instance, higher yields might allow farmers to intensify their production on existing farmland, leading to a reduced incentive to spare land for conservation.

Moreover, the prospect of greater profits resulting from yield increases can incentivise the conversion of additional land into farmland. The allure of increased financial returns may tempt farmers and landowners to expand agricultural operations into previously untouched or environmentally sensitive areas. This unintended consequence can undermine efforts to conserve and protect natural habitats. These criticisms underscore the importance of implementing comprehensive and well-designed land-sparing strategies. Careful planning, effective policy measures and strong incentives are required to ensure that yield gains translate into meaningful land conservation outcomes. By addressing potential rebound effects, perhaps with policy measures and by closely monitoring land-use changes, it might be possible to mitigate the risks associated with increased profitability and prevent the expansion of agricultural activities into ecologically valuable areas.[15]

While the concerns raised by critics should not be disregarded, it's crucial to approach them as opportunities for refinement and improvement rather than reasons to dismiss the land-sparing approach altogether. The two professors point to a recent reassessment of data for birds and trees in Ghana and India. The assessment showed that land-sparing would still outperform land-sharing even when policies to limit rebound effects are not put in place.

Bateman and Balmford have noted the concept of the 'focusing illusion', put forth by renowned psychologist and economist Daniel Kahneman.[15] This sheds light on a psychological phenomenon that affects our perception of the consequences of a change. It suggests that when we focus on a specific effect of a change, we tend to overlook or underestimate the multitude of other potential effects that may arise. In the context of agriculture and land-use decisions, this phenomenon can be particularly relevant. Often, studies concentrate on a single aspect or measure, such as biodiversity, carbon storage, nitrogen pollution, flooding, food production or recreation. While these single-focused evaluations provide valuable insights into specific outcomes, they may inadvertently disregard the broader range of impacts or fail to capture the systemic effects of a change.

By expanding our scope and adopting a holistic approach, we can better understand the complex interactions and trade-offs that occur within agricultural systems to help protect our forests and other ecosystems.

Toothpaste – I Once Was a Tree

Did you know that some toothpaste ingredients have a connection to trees? One such is xylitol, a natural sweetener commonly derived from tree bark, including birch trees. Xylitol is a sugar alcohol that gives toothpaste a pleasant taste and provides several oral health benefits. It's known to help prevent tooth decay and cavities. How does it work? Well, some bacteria in our mouths love to feed on sugar, and when they do, they produce acids that can attack our teeth. On the other hand, these bacteria cannot easily metabolise xylitol.

So, using toothpaste containing xylitol disrupts the ability of bacteria to thrive and produce harmful acids, ultimately promoting a healthier oral environment. Not only does xylitol offer dental benefits, but it also adds a touch of natural sweetness to our brushing experience. With each dollop of toothpaste on our brush, we have trees to thank for keeping our smiles bright and our mouths healthy.

CHAPTER 9

The Miyawaki Method

"We abuse land because we regard it as a commodity belonging to us. When we see land as a community to which we belong, we may begin to use it with love and respect."
—Aldo Leopold

I sat entranced as the freshly sprouted leaves shimmered in the spring sunlight. Those that had fallen the previous autumn were dancing in the breeze on the forest floor – a strange juxtaposition between this year's life and last year's death. Then I smiled, knowing that last year's fallen photosynthetic foliage was now partaking in another form of alchemy, breaking down and feeding the soil, microbes and trees: this year's life. I was in a forest of sorts – a tiny forest about the size of a tennis court, known as a Miyawaki forest.

Miyawaki forests are named after the late Japanese botanist Akira Miyawaki (1928–2021).[1] Miyawaki was an expert in plant ecology who specialised in natural forest restoration, particularly on soils that had been highly degraded. During a stint as a research assistant in an ecology lab in Germany, Miyawaki worked on a concept called Potential Natural Vegetation.[2] This refers to the vegetation assemblages expected to occur in an area or 'ecospace' without human intervention or major hazardous events. After returning to Japan in 1960, Miyawaki began mapping the potential natural vegetation of his homeland, using it to carefully select tree species for reforestation projects. He then created a large seed bank, often collecting seeds from remnants of natural forests that were scattered around cemeteries and temples. These trees were typically preserved owing to traditional beliefs in 'chinju no mori'.[3] This describes the trees around shrines that protected sanctity; it was considered unlucky to interfere with them. This belief enabled the preservation of thousands of tree species, their seeds and genes, all descended from ancient forests that grew naturally in bygone eras. Using this method combined with sourcing locally native seeds, Miyawaki began to run field trials to test his first tiny forests. He demonstrated that they grew quicker and were generally more resilient than typical monoculture plantations that failed to replicate natural conditions.

Delighted by his findings, Miyawaki proposed a plan to restore native forests across Japan. He told people his method of rapid forest restoration could enhance biodiversity and water retention and protect against natural hazards. However, his initial proposals were not always met with open arms, with some criticising the planting density, 'forced photosynthesis' and lack of evidence. But despite this initial scepticism, in the early 1970s a large steel corporation became interested in Miyawaki's work after its previous conventional plantations died, and he was hired to plant forests on its site. Miyawaki identified the potential natural vegetation of the area and investigated the remnant forests of surrounding cemeteries. He selected various species of locally native trees and created a nursery to cultivate the trees before planting them in his characteristic dense formation. Unlike the conventional plantations, the tiny forests were a success. The steel corporation commissioned a series of tiny forests at its steel mills across Japan.

Miyawaki and his collaborators developed and refined the tiny forest method and have since covered more than 1,300 sites.[4] The tiny forests are multi-layered and composed almost entirely of native species. They have been reported as successful in nearly all of Japan and many other places.[5] In Japan, Miyawaki forests are often used to boost biodiversity and mitigate the effects of tsunamis on the coastlines.

Miyawaki method in focus

1. The first step is to identify the native vegetation best suited to the area being planted (potential natural vegetation). This should involve intensive research by ecologists and members of the local community. In natural conditions, forests are multi-layered. The Miyawaki method aims to mirror natural

conditions by planting four layers of vegetation. Up to 40 different species are selected to create balance and maximise density.[6]

2. The second step is to collect local seeds and create a nursery to germinate the seeds. Seedlings are then cultivated. Ideally, seedlings should be grown under shaded conditions and then acclimatised to the recipient environments before planting out.
3. The third step is to prepare the soil. Since Miyawaki forests are typically planted in nature-deprived areas, the soil is often degraded, compacted or waterlogged. It's therefore vital to ensure the soil has the appropriate aggregation, moisture, organic matter and microbial communities to support the forest ecosystem. This might involve bringing in local organic matter and even inoculating the soil with beneficial mycorrhiza. A 'compost tea' is also sometimes added. This contains fungi and fuel, such as liquid seaweed, to help the fungi grow.
4. The fourth step is to plant the saplings, typically with the help of the local community. Three to four tree saplings per square metre are planted in a random manner. This is supposed to mirror the way natural forests grow. The young saplings then start to form symbiotic relationships with the mycorrhizal fungi in the soil.
5. The fifth step is to heavily apply mulch using a straw-like material. This is thought to protect the soil, help it to retain moisture and reduce the number of weeds. Many soil creatures draw down the organic matter, helping to build soil fertility and structure.
6. The final step involves management for the first couple of years. This may include watering, weeding and mulching when needed. After this, the idea is to let the forest take care of itself – a self-sustaining tiny forest ecosystem.

I recently heard about a Miyawaki forest created in a village called Witney near Oxford, which was planted in 2020 by the local community (the Witney Land Army); it's widely marketed as the "UK's first Miyawaki forest".[7] Six hundred native trees and shrubs fill a 200 m^2 plot next to a sports field. I visited Witney on a nice sunny

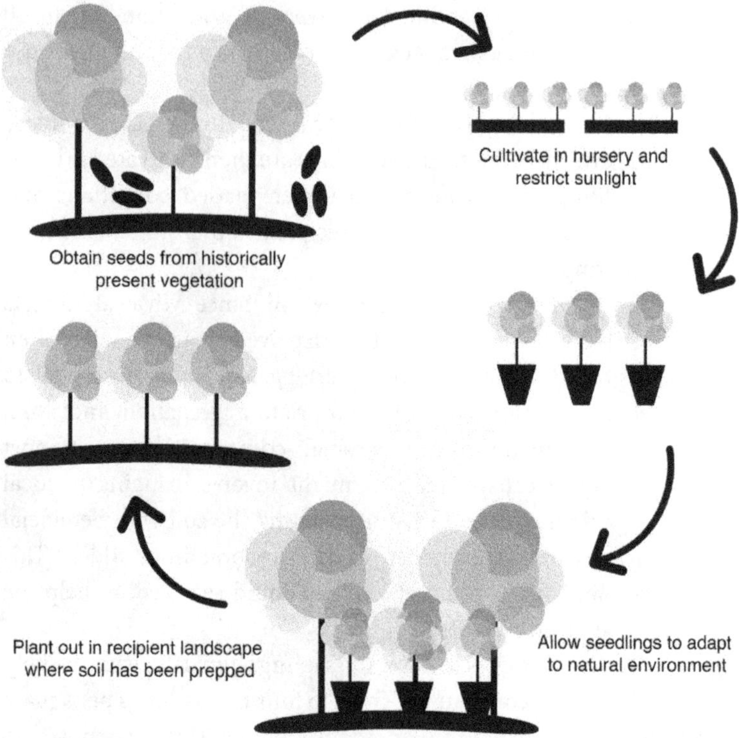

The Miyawaki method.

day in the spring. When I arrived at the tiny forest, I came across a cosy nook adorned with wooden logs, with visitors invited to take a seat and immerse themselves in the tranquil surroundings. This was referred to as the 'classroom area' on a notice board. I thought this was a great idea: it ensured the community, especially children, could engage with the forest, which is essential for fostering stewardship. As discussed in previous chapters, forests can be viewed as socio-ecological systems, and it's important to involve the local community to encourage reciprocity and healthier connections between people and nature in these forests.

After just five minutes of walking around the tiny forest with my ecologist's hat on, I was pleasantly surprised to find the number of species was fairly high. The botanical diversity included dogwood, hazel, silver birch, pedunculate oak, goat willow, guelder-rose and a healthy sprinkling of herbaceous species. Invertebrates buzzed all around. Reptile refugia (small square mats) had been laid down in the open patches of the tiny forest. Underneath them were earthworms, slugs, centipedes and ants. I spotted an adorable dark-edged bee-fly hovering and swiftly darting between the birch branches. Some describe bee-flies as tiny, fluffy narwhals of the sky. Their bodies are covered in hair, and they have a long proboscis (a tubular sucking mouthpart) for lapping up the nectar in flowers: the one I saw had a proboscis that was as long as its thorax and abdomen combined. Incidentally, part of the bee-fly's lifecycle is parasitic. The adult female flicks her eggs into the nest of solitary bees when the bees are away from the nest. The bee-fly larvae then feed on the bee larvae. Despite this rather morbid phenomenon, they don't harm the overall population of bees. They also parasitise wasps that parasitise spiders. So, they're a parasite of a parasite. The bee-flies are invaluable pollinators, helping the forest vegetation to flourish.

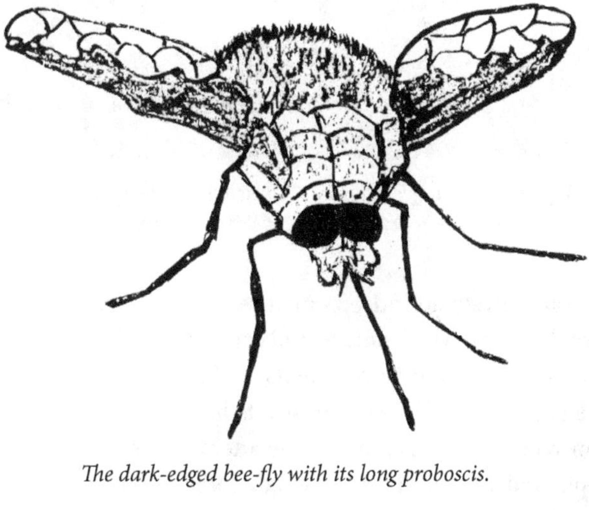

The dark-edged bee-fly with its long proboscis.

I was also treated to the chorus of passerine birds fervently competing for mates as the breeding season was well and truly under way. Great, coal and blue tits darted in and out of the tiny forest, picking up insect food and nesting material. Blackbirds perched on the top of the silver birch trees and a 'charm' of goldfinches swooped in, flashing their vivid gold and red feathers. I counted nearly 50 species of plants, birds and insects in just a few minutes. Although this is just one example, it supports the many people who say Miyawaki forests are great for bio-diversity. Only time will tell whether the Witney forest (or Tychwood, as it's known, cheekily referencing the 500-ha Wychwood Forest just up the road) will remain a mini-nature haven in the long term.

Witney's Miyawaki forest – you can see the height of the two-year birch tree growth in relation to the fence (approximately 1 m high).

If Miyawaki forests do indeed enhance local biodiversity, this would be a clear benefit – what some would call a 'nature-based solution'. But what are the other perceived benefits, and are there any disbenefits?

The popularity of Miyawaki forests has experienced a noticeable surge in recent years, capturing the attention of a growing number of people and organisations. Indeed, some people view them as a panacea or 'silver bullet' in relation to the climate and biodiversity crises. Yet, despite the potential hyperbole, the popularity of these tiny forests can be attributed to their results. Many people see the before and after photos of these forests, and the observed transformation is not typical of standard reforestation methods. Some of the purported benefits include cleaning the air, cooling the concrete jungle, enhancing habitat for wildlife and sequestering carbon.[8] However, critics argue that the theory and practice are ahead of the evidence. They also note that the Miyawaki method is expensive and the long-term benefits are not proven.[9] Based on the evidence I've read in scientific papers, my initial view is that Miyawaki forests can bring a range of benefits in specific ecosystems. But there's one caveat: they must be done in the right way.

As discussed, planting (and caring for) trees can bring huge benefits if it's done correctly. In addition to the environmental benefits, tree planting can have social benefits, such as enhancing social cohesion and employment opportunities. But planting the 'wrong' trees or using the 'wrong' soils (or planting on land that wasn't previously a forest ecosystem) can be harmful to the ecosystem and hamper resilience. However, owing to the rapid spread of the Miyawaki method globally, many tiny forests have been created without the proper foundations. In other words, trees have been planted under the guise of the Miyawaki method without sourcing appropriate local natural vegetation and making sure soil conditions are suitable. This puts a black mark against Miyawaki's tiny forest concept.

Miyawaki forests in India

An example of this malpractice has been reported in Chennai (earlier known as Madras), India.[10] Chennai is the capital city of Tamil Nadu state in southern India and has a tropical wet and dry climate. In the past, Chennai's natural vegetation was likely dominated by secondary vegetation in the form of woody thickets, cactus-like euphorbia and palmyra palms. The intervening spaces would have been covered in herbs and grasses. Trees were sparse, and the environment was likely to be semi-deserted. This is supported by the persistence of a handful of animals adapted to semi-desert habitats – such as the fan-throated lizard, blackbuck, saw-scaled viper and the Indian stone curlew. These animals survive in small, fragmented areas of semi-desert habitat in and around Chennai today. There were also probably mangrove trees along the coastline until around 100 years ago.

Planting trees in the name of green development is becoming increasingly popular. However, if planted in landscapes with historically low tree biomass, this can be dangerous. Unfortunately, it appears that this is playing out in Chennai, where the over-burdening of the landscape with trees is causing adverse impacts. R.J. Ranjit Daniels of the Care Earth Trust has said that the tree burden has almost exterminated the blackbuck, a species of deer. "Excessive shading and the micro-habitats created by trees could have been responsible for the rapid spread of the *Aedes* mosquito (a malaria-carrying vector). Their relatively poor transpiration combined with their dense growth and resultant solid waste accumulation causes water stagnation for months after the rains."[11] I spoke with Daniels via email, and he summarised his experience of the Miyawaki approach as follows: "Miyawaki forests are the opposites of bonsai, where trees are planted close to each other and forced to grow into poles as in certain types of plantations. The ecosystem services provided by these forests have not been scientifically studied. Blind replication of an alien system of mixed-species tree plantation could serve very little ecological purpose." Clearly, Daniels is not a supporter of the Miyawaki forests, and it seems this may be justified in the context of Chennai.

Leading the green development charge in Chennai are proponents of the Miyawaki method – often volunteers who work for small NGOs. According to Pradeep Tripathi, founder of an NGO that plants trees in Chennai using the Miyawaki method, many tree-planting groups are untrained in the method. This, he says, is tarnishing the Miyawaki concept. In an article written by Nikhil Eapen for *The Hindu* (one of Chennai's leading news outlets), Tripathi says, "The majority of groups don't conduct surveys to find out what the native species of the landscape are. They plant whatever they find."

This approach risks the local ecosystem and may increase the chances of creating monoculture habitats. Daniels is even more sceptical. He says that even if the Miyawaki method is done correctly, there have been no rigorous assessments to compare the approach with other restoration methods in India, especially in terms of how they affect biodiversity and the functioning of the ecosystem. Long-term studies are also scarce. Miyawaki forests may bring considerable benefits in the short term, but how do they stack up against traditional reforestation methods in the long term? These are all questions that researchers need to tackle. I decided to see if the Miyawaki forests peppered around the Chennai landscape had public reviews on Google. After all, they're a new landscape feature and would probably attract attention from local communities. I wondered if they would be positive or negative – an anecdotal assessment of public perception. I went to Google and typed "Miyawaki forests, Chennai" into the search bar, and voila! At least 15 different Miyawaki forests popped up on the map, all with public reviews. I'm not sure what I was expecting, but the general perception seemed overwhelmingly positive, ranging from 4.1 to 5.0. Here are some of the comments left by members of the public in Chennai:

> Peaceful place and greenery that's an ecosystem of its own.
>
> This forest going to be one of the best, and it's a good learning experience for the students who are studying there. Thanks to Vidaigal Thannarva Amaippu, Sastha Helping Hands and NSS students of Adhi Engineering College.
>
> Wow!!! From the railway station, what a fresh breeze from the greenery forest cover.
>
> Mini forest to make us feel connected to nature.
>
> Forest that will provide a habitat for many birds.

From this very rudimentary assessment (and it should be noted that no negative comments were left), it appears that people enjoy the tiny forests. They bring peace, comfort, perceived biodiversity and educational benefits. These factors should be taken into consideration when planning Miyawaki forests. Public benefits, to health and wellbeing, for example, and public support will be vital to their long-term success. However, perhaps an equally pertinent comment is this one:

> Great initiative. Maintaining such vegetation in a place like Pallavaram (in Chennai) is not easy. It's a result of serious efforts by good-hearted volunteers.

This points to the concerns about sourcing appropriate natural vegetation in a given area.

Miyawaki in Europe

Researchers at Wageningen University in the Netherlands conducted a study on new Miyawaki forests. They concluded that the tiny forests "increase the biodiversity compared to the nearby forest. Both the number of species groups and the number of individuals is generally higher than in the reference forests."[12] Fabrice Ottburg, the ecologist who led the study, thinks the enhanced biodiversity value may be partly attributed to the high number of plant species that provide various habitat features for foraging, breeding and sheltering for different creatures such as birds, insects and amphibians. In the early stages of a Miyawaki forest, the openness allows sunlight to reach flowering plants that seduce pollinators with their colours and pheromones, thereby boosting local biodiversity. Advocates say Miyawaki forests can grow 10 times faster and become up to 100 times more biodiverse than those planted by conventional methods. There are also (non-peer-reviewed) reports of Miyawaki forests having the capacity to store 40 times more carbon than single-species plantations.[13] In the Netherlands, the conservation group IVN Nature Education helped cities plant more than 200 tiny forests by the end of 2022.[14] Support by ecologists for Miyawaki forests seems stronger in Europe than in India. Although I've heard some critical comments from ecologists on social media, this general support may indicate an unhealthy dose of naivety. But it may also be that the conditions for Miyawaki forests are better in temperate landscapes with appropriate potential natural vegetation.

Criticism of the potential natural vegetation approach

It's important to note that the concept of potential natural vegetation is subject to debate. This means the concept that underpins the Miyawaki method is questionable, and therefore that the Miyawaki method itself should be scrutinised accordingly. Potential natural vegetation is debated on similar grounds as the climax theory, whereby vegetation in a community undergoes successional development until an optimum sustainable community (a climax community) in equilibrium with its environment is reached. Critics of the potential natural vegetation concept argue that ecosystems are ever-dynamic; that is, over time, vegetation communities are not static assemblages. Bioclimatic conditions constantly evolve, so it's illusory to define either a primary or a final vegetation stage in a given ecosystem.[15] Moreover, potential natural vegetation is supposed to be the community that existed before significant human interference. However, several studies have highlighted problems with this definition. For instance, using pollen records, one study found that before human settlement, the forests in La Laguna, Tenerife, were dominated by the genera *Quercus* (oak) and *Carpinus* (hornbeam).[16] These are currently considered non-native to this archipelago. Some researchers have said this example falsifies the expectation that potential natural vegetation in this archipelago is a community dominated by laurel forests.

In an ever-evolving world, where dramatic changes to the climate are currently being felt and are predicted to intensify, is it realistic to create new habitats based on historical conditions? Taking everything into consideration, it appears that Miyawaki forests can hold considerable value, particularly in the short term, when implemented properly and planted in suitable locations, especially urban areas. They can enhance local biodiversity and human wellbeing and have the potential to store large amounts of carbon. However, long-term impacts have not been sufficiently studied, mostly because the concept is only a few decades old. Let's for a moment consider Miyawaki forests in the context of replacing standard monoculture plantations or restoring degraded land in cities that have endured significant human disturbance. In this instance, I think there's a reasonable argument to support them. However, as discussed earlier, they can put pressure on certain sensitive ecosystems, and many questions remain unanswered and require further research. Moreover, as I've argued in previous chapters, natural forest regeneration is probably the most favourable form of forest restoration on a wider scale – so long as it doesn't exclude local human communities. Developing a set of Miyawaki forest standards and/or an accreditation system for forest planting operatives may be worthwhile. First, it would mean people could adequately assess whether the Miyawaki method can achieve its aims in different contexts. Secondly, it would mean people could fully understand the potential for any major unintended consequences and prevent them from happening. Adversity in this sense could play out in one of two ways. First, the initial problem (a lack of forest habitats) is addressed, but there are unintended consequences of the Miyawaki approach that result in adverse outcomes. Secondly, the initial problem isn't solved, and the Miyawaki concept worsens the situation.

There's a Miyawaki forest revolution taking place. Only time will tell if it will be a game-changer or a distraction from better solutions.

Aromatherapy oils – I Once Was a Tree

When it comes to the soothing scents of aromatherapy, it's intriguing to discover that many essential oils used in this practice have roots in tree bark and leaves. Take, for example, the invigorating aroma of pine and cedarwood oils. These oils are derived from the bark or leaves of pine and cedar trees, known for their rich and distinctive fragrance. Extracting essential oils involves carefully steam distilling or cold pressing these tree parts to capture their aromatic compounds. Once extracted, these concentrated oils are used in aromatherapy to promote relaxation, enhance mood and provide therapeutic benefits.

The scent of pine oil, reminiscent of a peaceful forest, is often associated with feelings of calmness and rejuvenation. With its warm and woody aroma, cedarwood oil is believed to have grounding properties that can help relieve stress and anxiety. So when you inhale the comforting scent of pine or cedarwood oils, spend a moment to appreciate the connection to nature and the power of these tree-derived oils in enhancing your wellbeing.

PART III

FOREST RESTORATION: A CONTROVERSIAL BUSINESS

CHAPTER 10

The Trillion Trees Controversy

"What we are doing to the forests of the world is but a mirror reflection of what we are doing to ourselves and to one another."
—Mahatma Gandhi

In 1977, Wangari Maathai could see that the top was 'heavy', and that changing those at the top posed a great challenge. But changing the top without a 'grassroots movement' was almost impossible. Rather than relying on beneficial actions to 'trickle down', Maathai advocated for initiating change at grassroots level within communities, thereby sparking a 'trickle-up' movement.

The movement I'm referring to is the Green Belt Movement – an environmental, grassroots NGO based in Nairobi, Kenya.[1] The movement focused on planting trees, environmental conservation and women's rights. It involved planting millions of trees to reduce the effects of deforestation while providing a forum for women to be leaders. Crucially, the planters were educated on the physical demands of new seedlings and the importance of nurturing trees to ensure their survival. This is the difference between planting and growing: the former is the mere act of dropping a seedling in a hole, while the latter represents a commitment to life, caring and the unfolding of potential.

The Green Belt Movement eventually stimulated the Billion Trees campaign, launched in 2006 in response to the challenges of climate change.[2] Subsequently, in 2011, discussions surfaced around launching a Trillion Trees campaign. This ambitious initiative has been championed by multiple organisations since then, such as Plant-for-the-Planet, BirdLife International, Wildlife Conservation Society and the World Wide Fund for Nature, propelling the campaign onto an international stage.[3] In 2020, the World Economic Forum also supported the Trillion Trees initiative, with messages such as "stop talking, start planting".[4] The campaign was in part kickstarted by a 2019 publication in the prestigious journal *Science*,[5] but the approach taken in this study was questionable, as was the communication of the results.

The *Science* article

First, it's unlikely that the researchers who wrote the *Science* article were being intentionally deceptive. I truly believe their general aim was and is to help restore global ecosystems and conserve biodiversity. It was probably a matter of poor communication, a couple of data oversights and a splash of naivety. In some people's eyes, it was unfortunate that the study helped to fuel a tree-planting movement when a tree growing movement was needed. It also helped fuel the idea that we must plant trees everywhere to 'save the planet'.

What were the study's outcomes, and why have some of its findings sparked contention? According to the authors, there was enough available space on the Earth's surface for an additional trillion trees. Moreover, the article suggested these trees had the potential to capture more than a third of the greenhouse gases released by human activities since the beginning of the industrial revolution, amounting to approximately 200 gigatonnes of carbon.

As reported in a *Nature* article, during an interview with NBC News, one of the leading scientists behind the study reiterated the oft-repeated phrase "stop talking and start planting".[6] You can imagine the subsequent media blitz that rallied excitement for this seemingly easy-to-implement climate solution. Headlines in newspapers, magazines and online news sites worldwide declared mass tree planting as the ideal solution to the climate crisis.

As the saying goes, "Problems can be complex, but solutions can be simple". Plant some trees and voila, CO_2 in the air will disappear, and everyone can relax. These headlines were, of course, hyperbolic and misguided. Following the media blitz, a group of 46 scientists responded to the study with their critiques. The lead author of the *Science* study rebuttal, Joseph Veldman, argued that planting trees where they don't belong can harm the environment and even exacerbate climate change – the very thing planting trees was supposed to solve.[7] The rebuttal committee argued that the amount of carbon that a trillion trees could sequester was exaggerated by five times in the *Science* study. It included savannah and grassland in its estimate of land suitability for tree planting, which some researchers claim means the study overstated where new forests could grow. Others thought the study might be used to justify indiscriminate tree planting, with potentially damaging consequences for biodiversity and people. It's important to note that the research lab behind the study has lamented the miscommunication, with the lab leader calling it the "greatest failure of my career".[8] The lab seems dedicated to helping to solve the world's restoration dilemmas.

Forrest Fleischman

Forrest Fleischman is an academic who researches environmental politics and forest policy, particularly in India and Mexico: he was one of the authors who responded to the *Science* article. When I caught up with him to talk about the article, he said, "We repeated some of the analysis in the study but with more accurate variables and found this trillion trees figure is five times too much! … If you exclude the landscapes in higher latitudes or where the trees don't grow naturally [e.g., grasslands, which are ecologically valuable in their own right], it was more like 200 billion. And we're not sure about this figure; it's possibly still too high, but we were making a point."

Forrest objects to the current thinking that we can simply put vast swathes of trees on the land to reduce the impacts of climate change. "Well, you know what? Something else is happening on this land right now. Degraded land is often a political category, not an ecological one. In the mid-west [USA], the land is covered in cornfields, which used to be forests and prairies. But it's productive. There's a reason we don't want to plant trees on it. The same goes for many other places in the world. The land is used to support people's livelihoods. Those leading these campaigns may think more people will be harmed by climate change than the few people working the land with goats."

Forrest believes we should explicitly discuss the potential impacts on different social groups when discussing tree planting. Displacing people who contribute far less to global climate change may be one of these impacts. "We should tackle the broader issues; that is, rich Western people contribute way more to climate change than Indigenous Peoples in the global south." Forrest is concerned that the *Science* article and the Trillion Trees campaign will drive indiscriminate tree-planting movements, displacing Indigenous Peoples and harming ecosystems. "These days, everyone seems to think that planting trees is a vital solution to the climate crisis. They're often wrong. Instead of planting trees, we need to talk about people sustainably managing landscapes."

The ten pitfalls of tree planting

Forrest and a team of other scientists recently published an article describing ten pitfalls of tree planting.[9] The first pitfall is that ecosystems, not tree-planting campaigns, capture and store carbon. "Tree-planting campaigns have high failure rates, and many ecosystems with sparse tree cover store large amounts of carbon below the ground."

The second pitfall is that protecting ecosystems, not planting trees, is the most cost-effective natural climate solution. Forrest suggests that tree-planting campaigns divert funding from this low-cost priority and fail because they do not address the drivers of forest loss. The main driver of forest loss is export-oriented commodities. "To protect and store carbon in ecosystems, we need to secure the property rights of people who have incentives to conserve that carbon – often Indigenous or otherwise resource-dependent people in the global south."

The third pitfall, Forrest maintains, is that tree planting is often unnecessary to restore ecosystems. In many situations, trees can grow naturally without anyone planting them, and planting the wrong trees in the wrong places can hinder regrowth. When restoration does involve planting trees, it often works better to plant a small number of trees that target specific goals. These can include establishing seed sources or making the new forest more economically valuable to local users than a large-scale planting programme.

The fourth pitfall is that tree plantations sequester less carbon and less securely than naturally regenerated forests. Widely used plantation practices result in forests with little long-term carbon storage and emphasise species that increase fire risk.

The fifth pitfall is that planting trees in grasslands, shrublands and peatlands can destroy biodiversity. "For example, global restoration plans or the Trillion Trees campaign would destroy the iconic African savannahs, replacing them with large-scale plantations."

The sixth pitfall is that trees can reduce water availability. "Yes, trees improve ground and surface water recharge, but they also use a lot of water. The balance between these is complicated and can only be determined in a site-specific fashion."

The seventh pitfall is that trees in the wrong places actually warm the atmosphere more than their carbon storage contributes to cooling. This is because trees are dark – so they absorb more heat than surfaces such as snow or dry grass.

The eighth pitfall is that "perverse financial incentives lead to rushed planting and high tree mortality. Hardly a week passes where I don't read a story about a tree-planting drive which was rushed for publicity and no allocation was made for maintenance. The trees are dead."

"The ninth pitfall is that tree planting can threaten rural livelihoods. Three hundred million people live on land targeted for restoration, and many depend on subsistence agriculture or grazing animals – land uses that often conflict with restoration."

Finally, the tenth pitfall is that tree planting targets the Global South to capture emissions from the Global North, raising fundamental questions of environmental justice. If tree planting is beneficial or benign, there may be no problem. But, as Forrest says, "We show that tree planting can disrupt the livelihoods of hundreds of millions of the world's poorest people while also destroying biodiversity and undermining the provision of ecosystem services like water."

This may all seem like doom and gloom. Yet Forrest says we're just doing things wrong, and his work presents an opportunity to get things right. He suggests that instead of focusing on tree planting as a "natural climate solution", we must focus on "people-centred climate solutions". This means supporting social systems that encourage people to conserve ecosystems. "Long-term investments in tree growing require secure land tenure for rural and Indigenous Peoples, incentives and access to investment capital. These are lacking for many in the Global South and should be a priority for those who care about restoration."

Some people on social media have disagreed with Forrest on some of these points. One person said, "I'm more inclined to say that tree planting is one solution, with pros and cons in balance. You focus on cons here, but I agree with you that ecosystem protection should be a much bigger focus than it is now." Forrest thinks that tree planting is never a solution. Tree growing in the context of a more extensive set of policies that support long-term carbon storage is a solution, "and sometimes (but not always), you need to plant the trees you want to grow". Tree planting versus tree growing may seem like a semantic issue, but the implications of the wording are considerable. "Right now, we're seeing policymakers plan to plant trees and not think about why, where, or the policies needed to support tree growing in the long run." This doesn't mean that all tree planting is a bad thing. It's a generalisation for sure. Every year, millions of pounds are spent on attempting to restore landscapes by planting trees. Yet, despite this huge sum of money, restoration success is sporadic at best.

Researchers at World Agroforestry (ICRAF) say we need to rethink our words, as words are a powerful tool for shaping perceptions, perspectives and, ultimately, actions.[10] They point out that seedlings in mass tree-planting projects have a low survival rate. This makes for a poor return on investment and mocks myriad well-intended attempts to restore ecosystems. ICRAF thinks tree planting is too often stated as the ultimate objective in restoration projects. "We're going to plant a trillion trees", "stop talking and start planting" and "We'll combat climate change by planting, planting, planting!" They say the objective should be growing trees instead. Indeed, the success of landscape restoration projects is often measured by the number of trees planted. Some restoration projects operate on short timescales. They may have one to three years to play with, but many trees require five years or more to survive early threats such as adverse weather, grazing and disease. They often need to be cared for in their infant years. Either that or they need to have the right conditions to thrive naturally.

ICRAF engineer Judith Nzyoka recently wrote in an article, "Species should be selected that match not only the biophysical but also the socio-economic conditions."[11] Judith is not only acknowledging the need to get things right ecologically, but she's also pointing to the need to include people. Rural communities in areas under restoration are far more likely to care for the trees and help them grow if they are useful to them and if the local people are allowed to lead the process. Newly planted trees in degraded landscapes where humans live often require irrigation and protection against damage from fire and livestock. Moreover, as the climate is changing, and quite dramatically in some areas of the world, while tree species must match the local conditions, the predicted climatic impacts must also be considered and planned for accordingly. There's no point in planting something that can't adapt to future conditions. Essentially, Forrest and the ICRAF folk think restoration efforts will continue to fail without a proper engagement process wherein the needs of local communities are embraced.

Engaging with people in areas where people live is vital for forest restoration. But there are many places where forests can regenerate naturally without the need for significant human intervention. Forrest also recognises this notion. He told me how he grew up in New England in the USA, and his family had a modest patch of land with a cabin near a woodland. Forrest recalls one of his chores was to keep the land immediately around the cabin clear of trees and shrubs. "It was so hard to prevent the land around the cabin from turning into a dense forest. The main 'weed' was maple! You would clear the land of shrubs, and before you know it, the maple trees had recolonised." If the conditions are right, nature will find a way to restore itself.

Improving forest restoration

I asked Forrest, "What are the top three things we need to do to improve forest restoration?" His reply can be summarised:

1. We must avoid promoting all kinds of large-scale big target initiatives that ask, "Why are there no forests in those places", particularly where humans reside. We should focus on deforested areas, but first we need to evaluate whether people are currently using them for other things.
2. Empowering local people to restore and promote their land rights is vital. This is still not recognised enough by policy-makers. The most significant and impactful policy tool is promoting Indigenous land tenure – collective land tenure, where people have the right to decide how to manage the land sustainably. Currently, the land in many areas of the world is cleared for over-consumers. People are then displaced to make way for tree-planting schemes.
3. The science is clear on what works and what doesn't work. Yet people ignore it because it comes from the social sciences! We know that wealthy people's overconsumption drives forest

> degradation. As always, we need to address the root cause of the issue. We need to stop the patterns of wealthy consumers.

Forrest described an article in the German news source *Zeit* titled "Fairy-Tale Forests".[12] It talked about a tree-planting initiative in Mexico led by the NGO Plant-for-the-Planet. The organisation acquired four large areas in the Yucatán region of Mexico, described as "22,500 hectares of destroyed rainforest". However, the article highlights many inconsistencies in the project. According to Forrest, the region has "had incredibly stable and dense forests for decades compared to other parts of the tropics". Satellite images apparently show that the land Plant-for-the-Planet had acquired was not destroyed after all – the forest was intact. Therefore, reforesting the land makes no sense.

There are also reports that inappropriate planting between established trees and planting in flood-prone areas was happening in the Yucatán region. What prompts Plant-for-the-Planet to accept millions of pounds from benefactors, only to plant trees in already established forests where their survival is challenged or in regions that are evidently prone to disaster? Forrest reflected: "It's interesting because it's a huge failure, and one of the project leaders works for the same lab that published the *Science* article. In fact, Plant-for-the-Planet has strong ties with the lab. It illustrates that a very well-meaning group of smart Europeans have parachuted into Mexico to restore their forests but completely screwed it up. The locals knew how to restore their land, but the tree planters didn't engage with them properly or understand their land. They also planted where trees are not needed. This is an example of the neo-colonial mindset. Many Indigenous Peoples with traditional ecological knowledge are also trained foresters with university degrees. These people are being bypassed to make way for failing initiatives led by the Western community."

Economics and the right tree in the right place

Another factor to consider is cost. Tree planting can be costly if the wrong trees are planted in the wrong places. However, our governments still fail to fund proper tree-planting (or growing) efforts despite the fundamental importance of forest ecosystems to biodiversity and human society. According to the UK's Department for

Environment, Food, and Rural Affairs, less than £1 per person per year is spent on planting trees in England.[13] To put this into perspective, we can compare it with £90 per person per year on roads and £150 per person per year on fossil fuel subsidies.[14] These minimal efforts risk becoming even more futile if we neglect to comprehend the ecology of the trees, the characteristics of the land where they are planted and the communities that care for them.

Many have suggested that we all need to start thinking about growing the right tree in the right place to improve this situation. From an ecological perspective, let's go deeper than that and think about the tree as an ecosystem in itself. As I'll discuss later in the book, trees can be viewed as holobionts – a host plus trillions of microbial partners working symbiotically to form a functioning ecological unit. We risk the trees perishing if we plant without understanding their complex relationships with the soil microbiome. Perhaps instead of the right tree in the right place, we need to start thinking about planting the right system in the right system. A tree is a system in itself, the environment it's planted into is a system – and involving the latter is imperative to the forest's long-term future. As a society, we need to spend more money on forest restoration (as with other ecosystems). Still, even the current expenditure will be wasted if we fail to take a holistic approach.

Restoring forests in South Australia

"When we say 'woodland', in South Australia it's naturally sparse – only 20–30% of tree cover", Shaun Kennedy said to me.

I recently met Shaun, a revegetation expert working for SA Water in Australia. We spent a morning hopping barbed-wire fences and collecting soil samples for a research project and had a good natter about the local landscape. Shaun told me about the restoration successes and failures in the Mediterranean biome of South Australia. This area is characterised by a species-rich ground layer, shrubs with leathery leaves and a scattering of trees and scrub. Earlier restoration efforts in the area involved planting many trees with the aim of restoring complex forest ecosystems. However, this largely resulted in a monoculture habitat with a low number of species occupying the space between the ground and the canopy. The restoration folk were essentially creating a habitat that wasn't there before. This is why the definition of 'forest' or 'woodland' needs to be deeply considered before trying to restore it.

Picture a forest. What do you visualise? Many may imagine a crowded expanse of trees. But the natural landscape of this region in South Australia tells a different story. Here, the terrain is a tapestry of broad spreading trees over a diversity of shrubs and often open grassy meadows in a tangled mosaic. Yet the human-made forest was a sea of uniformity. And it wasn't just the flora that felt the impact. The wildlife found it difficult to return. Many birds and invertebrates relied on the naturally evolved Mediterranean vegetation, which, as mentioned, had a diverse ground layer providing a floriferous bounty of niches.

Shaun told me how the density of trees planted also affected birdlife in the canopy. Ordinarily, towering gum trees would rise, not in dense clusters, but as solitary giants scattered across the landscape. Their sprawling canopies created a unique habitat, a lofty realm where certain birds could set up their homes and forage for food. These spaces nestled among the leaves and branches were sanctuaries in the sky. However, the newly planted trees were so dense that they shot up towards the sun as they competed for light with their neighbours. The result? The spacious, sprawling canopies that once hosted a variety of bird life didn't exist and the birds often failed to return. The overly dense forest habitat only supports a narrow range of generalist bird species and appears to reduce the insect productivity that forms an important protein source for many of the region's honeyeaters.

As with many interventions in nature, this one carried a valuable lesson, and Shaun told me how they're thinking differently about the restoration process now. The phrase "Plant the right tree in the right place" perhaps requires an addendum: "in the right numbers and in the right way". He said this was also a top-down approach. It's an approach that seldom works to bring about the reassembly of rich understorey species.

Now, they're experimenting with planting the low-level grasses, forbs, wildflowers and shrubs first – a bottom-up approach. And they're having great results. One of the key lessons is that when we think of restoring a forest, we should first gain a deep understanding of what the ecosystem was like before it was degraded. Did it have a mass of crowded trees or was it characterised by a light scattering of trees with large open spaces filled with ground flora? Before the

European invasion, the open spaces in South Australia were often maintained by large herbivores (kangaroos) and by fires controlled by the Indigenous Peoples. To restore the ecosystem, then, we also need to consider if these factors are still present. We also need to create the conditions for the vegetation to flourish. Often, 'legacy effects' work against forest restoration projects – these are past actions or inputs that affect present ecological processes. An example of a legacy effect is the accumulated nutrients pumped into the ground by previous agricultural practices. The subterranean environment has been altered so much that simply sticking trees in the ground and expecting them to flourish is unrealistic. Leaving the environment to naturally regenerate might not work either because the natural seed bank may not exist, and invasive plants simply take over. It sure is a complex subject.

Overstocking

I've mentioned density a couple of times and how planting too many trees in one area can negatively impact the ecosystem. However, trees are sometimes planted at high densities, known as overstocking, to increase their survival chances. In some instances, this can be beneficial, but it depends on the local ecology, grazing pressure and management. For example, planting more saplings than are needed can allow some trees to survive if many get eaten by large herbivores or damaged by invertebrates or drought. If too many survive and the woodland becomes too crowded, they can be actively thinned out every few years – but this requires time and labour.

To summarise, while having additional trees on the planet (in the right places) would be fantastic, mass and indiscriminate planting campaigns are not the way to go. This approach often displaces Indigenous Peoples and harms ecosystems, and it has stimulated a vast greenwashing campaign. Switching our approach and language from planting to growing is vital. Planting is simply unnecessary in many cases. Mass tree-planting campaigns such as the Trillion Trees initiative also miss the point that forest restoration isn't always the most effective solution to tackling greenhouse gas emissions. Indeed, they may fail to acknowledge the importance of other ecosystems and address the root cause of the issue – the current patterns of wealthy consumers.

CHAPTER 11

Greenwashing Galore

"There are two ways to be fooled. One is to believe what isn't true; the other is to refuse to believe what is true."
—Søren Kierkegaard

A few years ago, I started seeing the term 'greenwashing' everywhere. It was environmentalist Jay Westerveld who first coined the term in 1986, to define the disinformation disseminated by an organisation to present its public image as environmentally responsible.[1] This was way before consumers used the internet and social media, and most people received their news from television, radio, newspapers and magazines. Back then, the combination of limited access to information and a seemingly endless bombardment of advertisements enabled corporations to present themselves as stewards of the environment, despite some of them engaging in environmentally unfriendly practices.

Around the same time as Westerveld came up with the term, oil giant Chevron commissioned a series of adverts to manipulate the public into thinking the company was environmentally conscious.[2] The adverts showed Chevron employees protecting kit foxes, bears, sea turtles and other cute animals. Here's an excerpt from one of their TV commercials:

> On a moonlit Californian desert, a [cute and cuddly] kit fox senses a prowling coati. Caught in the dangerous open, she speeds to the edge of an oil field toward one of these curious mounds specially made for her by the people who work there. She shoots through a pipe, too small for the

> coati, and into a cosy den that keeps her snug and safe ... Do people really do these things just to help an endangered species make it through the night? People do.

The commercials did what they aimed to do: fooled the public into thinking Chevron cared for the environment. The company even won advertising awards. Among environmentalists, however, they are considered the gold standard in greenwashing. Amazingly, the adverts cost millions of pounds to create, yet the conservation programmes described (which, I must emphasise, were mandated by law) likely cost in the thousands. Moreover, Chevron was considered the largest corporate polluter in the USA at the time.

Greenwashing as a concept dates back to earlier in the twentieth century, and probably before that. In the 1960s, nuclear organisations produced adverts to proclaim the cleanliness and positive environmental impact of nuclear power stations after a series of nuclear meltdowns across the USA, while conveniently ignoring the topic of nuclear waste. By the time the internet was publicly available in the 1990s, polls had shown that consumers were wising up to environmental sustainability issues and that a company's credentials influenced consumer purchases. By the end of the 1990s, 'greenwashing' was officially included in the Oxford English Dictionary.[3] A large poll involving millennials in the 2010s showed that over 70% of global consumers in that age group were willing to pay a premium for environmentally friendly products.[4] Many organisations profess to care for the environment, but consumers are becoming more aware of their greenwashing tactics. Another oil giant, BP, made intensive advertising and marketing efforts that fall into the greenwashing category before the US Gulf oil spill crisis in 2010. It's interesting how one can advertise a tremendous environmental consciousness one moment and the next contribute to one of the world's worst oil spills in recent history. Another form of greenwashing is to use a nature-based or green coloured label on something to make it appear more sustainable or beneficial to one's health. Coca-Cola did this in 2013 with their Coca-Cola Life drink.[5] The billions of plastic bottles they produce are far from environmentally friendly and the 6.6% sugar in the drink is far from being a health-promoting inclusion. You're more likely to reduce your "Life" if you drink it in abundance, as many people do.Innocent Drinks is another brand with a greenwashing past.[6] Innocent is owned by, you guessed it, Coca-Cola – one of the worst plastic polluters worldwide. Innocent produced a suite of TV cartoon adverts titled "Little Drinks, Big Dreams", featuring cuddly animals rejoicing in songs about recycling and caring for the planet. They claimed that buying their smoothie drinks could help save the environment. Contradictory, to say the least. Fortunately, the Advertising Standards Authority banned the ridiculous adverts after they were reported by Plastic Rebellion, an activist group. Their Matt Palmer recently said in the *Guardian*, "greenwashing is dangerous – in the case of Innocent, it's one thing to hide your ecocidal practices, that's bad enough, but to go to the next level and pretend you're 'fixing up the planet' is far worse. It means that people will willingly – and unwittingly – opt-in to support your project in the belief that they are doing good for the planet."[7] The bottled-water industry is equally pernicious, as it promotes images of mountain wilderness and pristine lakes to sell its products. The companies spend millions of pounds on advertising to convince the public that bottled water is somehow good for the body and great for the environment. In 2008, Nestlé Waters Canada even claimed that "bottled water is the most environmentally responsible consumer product in the world."[8] A global check of almost 500 websites found that four in ten used unsubstantiated claims or omitted information to appear more ecofriendly.[9]

How about trees and forests? How do these relate to greenwashing? Well, there's a tidal wave of organisations now starting to plant trees to make them seem more environmentally friendly and to keep consumers on their side. Unsurprisingly, many of us associate tree planting with beneficial outcomes for the planet, especially in countries such as the UK – one of the most nature-depleted countries in the world – so today a range of companies are using tree planting as a marketing tool. Trees are in vogue as attention on climate change and the biodiversity crisis increases. Businesses can now choose to offset their environmental impact by 'paying off' the damage through tree planting. But are they saving the trees or saving face?

While some people may argue that planting trees is a win–win for the planet, whatever the method or wherever the location, as I discussed in Chapter 10, planting and growing are two very different concepts. Offsetting is often a form of corporate greenwashing, and indiscriminate tree planting can cause more harm than good. For example, Eucalyptus is regularly chosen as a tree of choice in afforestation programmes owing to its fast-growing pace.[10] However, Eucalyptus is not native to many places but is now planted as part of offsetting schemes. It can out-compete native species in the recipient environment and drain the water table, negatively affecting the ecosystem.[11] Trees are also often planted in areas without consulting local communities. This may involve ignoring Indigenous land rights. Indiscriminate tree planting in these areas may exacerbate tensions between communities and authorities. For companies genuinely committed to mitigating their environmental impacts through tree-planting initiatives, it's imperative to prioritise the involvement of the local community in the planting process.

Another key issue with the tree-planting form of greenwashing is that there just isn't enough land on the planet to plant (or grow) enough trees to offset the carbon that the big polluting corporations are spewing into the air. Planting trees gives the impression that it's OK to pollute. If we were actually serious about transitioning away from fossil fuels, we'd be predicting that the carbon market would collapse. But the reason we can predict massive growth in demand for carbon credits (a tradable permit or certificate representing the right to emit one tonne of carbon dioxide as part of offset schemes) is that we're predicting enormous amounts of new sources of pollution. The more time we spend talking about carbon offsets, the less time we spend on decarbonising the economy.

Fossil fuels need to stay in the ground and renewable forms of energy need to become mainstream. A recent analysis by Oxfam suggested that we'd need at least 1.6 billion ha of new forests to achieve "net zero" by 2050.[12] This is the equivalent of five times the area of India or the combined area of Brazil and Australia. The report also found that the oil company Shell would need to reforest nearly 30 million ha of land using the tree-planting method to offset 35% of its emissions by 2050. That's the size of Italy.

While it's true that forest restoration is a vital part of mitigating the effects of climate change, that it's a panacea is a global political myth. Offsetting schemes were recently branded a "new form of colonialism" by Indigenous activists at COP26 (2021).[13] This is in response to their being exploited by big corporations to give the impression that they're ecofriendly.

A haunting truth

At the heart of these greenwashing campaigns lies a haunting truth. The term 'ghost acres' appeared in the 1960s.[14] It referred to land abroad that was – and still is – used to grow animal feed for the livestock we rear at home. The impact of our food consumption is obscured by complex supply chains and hidden in these ghost acres abroad. Yet the same phenomenon also applies within a country, for instance if we import compost or animal feed from another location in the same country. This can positively impact the recipient farm or garden by helping the ecosystem flourish while harming the land of origin. Out of sight, out of mind. If you spend a moment thinking about these ghosts, you may realise that we often choose not to see them. They're an 'inconvenient truth' but not invisible. A recent study suggested that a key driver of the deforestation of tropical forests is the pattern of over-consumption in other countries. Many of these countries are already nature-depleted owing to the same acts of land clearance that are now outsourced. The consumption patterns for G7 countries alone drive an average loss of four rainforest trees per person per year, an annual loss of more than 3 billion tropical trees just owing to G7 imports.[15] Other research has found that food and beverage consumption in 'rich' countries is the greatest driver of consumption-driven extinction risk – responsible for 39% of the global extinction-risk footprint.[16] This led to headlines such as "How your morning coffee could be driving species to extinction" and comments such as, "Buying coffee in Sydney could contribute to the loss of biodiversity in Honduras."[17]

How to spot greenwashing and what to do about it

We're bombarded daily with marketing spin, ploys and unsubstantiated claims that give organisations a false image of caring for the environment. But how do we spot and stop greenwashing? It's not always easy, but you can do a few things to make it easier. First, try to see through the sustainability buzzwords. To do this, you should look for evidence supporting any claims that an organisation is sustainable or has ecofriendly practices. You can do a web search to seek out a company's credentials or, more simply, look for marks and logos of approval from auditing organisations such as B Corporation or Fair Trade. B Corporation is a private certification of for-profit companies that's awarded once their social and environmental performance has been assessed. You don't have to stop at the logo, though. You can visit the B Corporation website, and when you type in the company name, it will provide you with a breakdown of their assessment and an overall score. If the company scores 80, they qualify for B Corp Certification. The median score for ordinary businesses completing the assessment is only 50. Amusingly, some companies advertise that they're 'B Corp Pending'. Some of these companies probably have honourable intentions, but I think this could also be greenwashing. A shiny 'B Corp Pending' sticker gives consumers the impression that the company is sustainable when it hasn't received the accreditation. Imagine putting 'X Executive job pending' on your CV when all you've got is an interview offer!

Secondly, try doing some research to assess a company's authenticity yourself. Ask yourself, "Does this company harm the environment in other ways? Are they merely planting trees to offset? What is their history regarding environmental issues?" Some organisations may have had an epiphany followed by a massive change in structure and vision. Still, more often than not, if it seems like greenwashing, the company's actions are probably trying to cover up something sinister. You could even send a message to the company asking for their policy on environmental issues.

Common sense trumps all. This may seem obvious but trust your gut instinct. If a company is doing something obviously detrimental to the environment and its policy is to plant some trees to 'offset' their emissions, it's greenwashing. If a company uses vague terms or environmental policies, that is, there's a lack of transparency, alarm bells should be ringing. A business genuinely working hard to minimise its environmental impact is usually happy to share its policy. If a company uses beautiful images of nature and it's hard to find its stance on environmental protection, it's usually best to steer clear – looks can be deceiving. If a company makes bold, scientific, 'green' claims without citing empirical studies, alarm bells should be ringing again. Unfortunately, some organisations use questionable scientific papers to support their cause, so this can be tricky to evaluate.

If you think you spot a greenwashing campaign, there are a couple of things you can do to tackle it. First, ask the company questions, take to social media and politely quiz it 'in public' – this alerts other consumers to the issues and any response will be seen by a wider audience. Secondly, don't buy into the organisation. If you spot a company employing greenwashing tactics, avoid it and try to seek out an alternative. Brands and manufacturers will get the message if their products and services become unpopular owing to greenwashing. And thirdly, if you spot a greenwashing campaign, the organisation may be breaking the law, particularly if it's making blatantly false environmental claims and misleading consumers. Many countries have agencies to deal with false advertising. In the UK, the Trading Standards and the Competition and Markets Authority can initiate court proceedings and the Advertising Standards Authority can take action – much as took place with the Innocent drink greenwashing adverts.

How neuroscience can lift the veil

Neuroscience research can be used to identify and overcome the threats of greenwashing. As discussed, greenwashing, the practice of making false or misleading environmental claims to promote a product or service as ecofriendly, has become increasingly prevalent. While consumers are becoming more environmentally conscious and seeking sustainable options, some companies are all too eager to capitalise on this trend. However, cognitive neuroscience can provide a powerful tool against it, to ensure that these false claims do not mislead us.

Cognitive neuroscience is the study of how the brain processes information and how this affects our behaviour.[18] It can provide insight into how consumers perceive and process information, including environmental claims made by companies. For instance, studies have shown that people are more likely to be influenced by information that's visually appealing or emotionally compelling. This means that companies can use images of nature or animals to make their products appear more ecofriendly, even if the product itself is not sustainable. Research has also shown that people tend to have a limited capacity for processing complex information and may rely on mental shortcuts or heuristics when making decisions. Companies can exploit this by using buzzwords such as 'sustainable' and 'green' without providing any meaningful information about a product's environmental impact.

Large corporations take advantage of the 'out of sight, out of mind' concept through greenwashing. Their marketing campaigns aim to take the public's eye away from the core issues, thereby putting them out of mind. To combat this, we can do several things, beginning with learning about cognitive biases, which are what can lead people to accept greenwashed claims without critical evaluation. Greenwashing exploits these biases, also known as cognitive blindspots. They are systematic errors in judgement and, as humans, we're prone to falling for these each and every day. One cognitive blindspot is the availability heuristic. This is the phenomenon whereby people tend to overestimate the importance of information available to them. It's a mental shortcut – the basic definition of a heuristic. The availability heuristic relies on immediate examples springing to mind when evaluating a particular concept or decision. Greenwashing campaigns exploit this by making only positive information available to consumers. Greenwashing is dangerous, and government agencies should do more to stop it.

The so-called halo effect is another example of a cognitive bias. It causes our impression of someone or something in one domain to influence our impression of them in other domains. It can lead consumers to over-generalise a company's green initiative and view the entire company as environmentally friendly. By being aware of the halo effect, we can counteract it. Simply knowing this bias exists and reflecting on it can prompt us to think more critically and not automatically assume that because a company is green in one area, it's green in all areas. To reinforce this, always ask questions about the company, look beyond single initiatives, research its other activities and consider the scale of impact.

Another neuroscience-based approach to overcoming the risks of greenwashing is to understand the persuasive power of emotion. Greenwashing often involves emotional appeals, such as guilt about environmental degradation or hope for a greener future. If we spend time understanding how these appeals affect the brain and influence decision-making, we can take a step back and critically evaluate the claims. We can also practise mindfulness techniques. This involves paying attention to your present experiences, including your emotions, in a non-judgemental way. By being more mindful, you're more likely to recognise when your emotions are being manipulated and make more rational decisions.

Cognitive reappraisal is a psychological technique in which the thoughts and assumptions that lead to emotional responses are challenged. For instance, if you feel guilt when a company claims its product is the only ecofriendly option, you might challenge this by considering other ways to be environmentally friendly or by researching whether the company's claim is true. Moreover, understanding framing effects can help. Framing refers to how information is presented, which can significantly influence our decisions. For example, a product might be framed as '95% ecofriendly' or 'only 5% harmful to the environment', which can elicit different emotional responses. Being aware of this can help you make more balanced decisions.

Remember, it's perfectly OK to have emotional responses to environmental issues – we need emotional engagement to motivate action. However, it's also important to combine this emotional engagement with critical thinking to avoid falling victim to greenwashing.

It's also possible for companies to draw upon cognitive neuroscience methods to understand how to present information in a way that's more effective at preventing greenwashing. For example, research has shown that people are more likely to be persuaded by information presented in a clear, concise and fact-based manner. Giving consumers accurate and transparent information about a product's environmental impact, such as its carbon footprint or the materials used in its production, can be more effective at preventing greenwashing than relying solely on visual cues. Cognitive neuroscience can also help companies to identify the most effective ways to communicate complex environmental issues to consumers. For example, research has shown that people are more likely to be influenced by stories or narratives that are emotionally engaging rather than dry statistics or facts.[19] This means that presenting consumers with real-world examples of how a product's environmental impact can affect people and the planet can be more effective at preventing greenwashing than simply presenting them with a list of statistics. Cognitive neuroscience can therefore provide a powerful tool for preventing greenwashing and ensuring that false environmental claims do not mislead consumers. By understanding how people process information and make decisions, we can develop more effective strategies for communicating accurate and transparent information about a product's environmental impact. Despite its risks, it can also help to promote sustainability and protect the planet for future generations.

Planting the right tree (or system) in the right place is vital for forest restoration. Still, many organisations exploit cognitive biases around tree planting to fool consumers into believing they care for the environment. Unfortunately, this has the effect of fuelling indiscriminate tree planting worldwide, which can be harmful to the environment and community livelihoods. Companies can exploit this concept via greenwashing, but we can do things to spot their tactics so we aren't affected by them. However, it's important to remember that the responsibility for preventing greenwashing should not fall solely on consumers. Companies should be held accountable for their environmental claims and practices, and regulations should be in place to prevent misleading marketing tactics.

CHAPTER 12

Sea of Tree Guards

"If lightning is the anger of the gods, then the gods are concerned mostly about trees."
—Lao Tzu

It was a warm day in spring, and I was heading to the West Country – an area of England famous for cider, Roman baths and Isambard Kingdom Brunel's Clifton suspension bridge. I didn't manage to see any of those things on my trip (although I did pass Stonehenge en route), for my destination was a grassy field in the middle of nowhere. There awaited Joe from the Woodland Trust – the UK's largest woodland conservation charity, and James, a local volunteer. The reason for my visit? To see and talk about a recent trial of 15 different tree guards – those ghastly plastic cylinders you see spread across the landscape – to test the effectiveness of non-plastic guards that purportedly 'biodegrade'. Joe and James took me for a walk around the field where 5,000 tree guards awaited in vast rows like a tiny plastic military tattoo, and explained how their trial began. Joe showed me each of the tree guards in the trial. There were some classic plastic ones for comparison joined by guards created from shiny cardboard, pulp, corn starch, cotton and pine oil, and wool and cashew nut oil. The Woodland Trust had recently worked with University College London (UCL) on a study on the environmental impact of protecting tree seedlings with plastic tree shelters.[1] I'll talk about this study later in the chapter, but essentially it showed that not using tree guards was the most favourable option across different environmental impact categories.

As we know, plastic waste is a massive problem globally. Owing to its chemical make-up, it can take over 400 years to degrade. Each year, we produce 380 million tonnes of plastic, and 8 million tonnes end

Tree guards in a field.

up in our oceans, with some suggesting this could double by 2025.[2] Visible chunks of plastic can harm wildlife by damaging their tissue and physiology. An equally important issue is microplastics. Before plastics can fully degrade, they first become microplastics, which are microscopic plastic particles less than 5 mm in size. These can harm wildlife. Some researchers have suggested they interrupt nutrient absorption, cause reproductive problems and hinder growth and survival.[3] A recent study showed that microplastics could induce stress, tissue and DNA damage in earthworms – vital members of the soil community.[4] Others say they can affect the soil microbiome (the microbial community) and even plant performance.[5]

There's a genuine concern that plastic tree guards – which aim to protect trees, thereby helping to restore degraded ecosystems – pose a considerable threat to biodiversity, thus being clearly counterproductive.

I spoke to the lead author of the UCL and Woodland Trust study, Charnett Chau, a biochemical engineer. A couple of years ago, the Woodland Trust approached Charnett's lab. It wanted help in reaching 'net zero' by planting 2 billion trees by 2050, which would supposedly increase the UK's tree cover from 13% to 19%.[6] The team was asked to evaluate the ecofriendliness of plastic tree guards, and Charnett's team found that, after considering the material type, the carbon emissions from production and the installation logistics, having no tree guards and planting more trees had the lowest impact when it came to climate change. The estimated survival rate of trees with guards was

85% compared to 50% without guards. So, for every two trees planted without a guard, one would not survive, but this still outweighed the disbenefits from the plastic and carbon emissions.

Charnett told me about the concept of 'incineration with energy recovery', which is exactly how it sounds: 41% of waste in the UK is incinerated and energy is recovered. Some waste companies use this energy to help power their own plants. In total, 16% of waste is incinerated without energy recovery, and the rest is sent to landfills. She continued, "Another thing to consider is that conventional plastics are derived from fossil fuels, so when we don't recycle them, and they end up being incinerated, carbon is emitted." On one occasion, she went to an old landfill site in East Tilbury by the Thames, which was active between the 1930s and 1980s. Over the years, the landfill has been exposed, and she got to see all the waste layers – newspapers, glass bottles, plastics all still there, including fragments where she could still read the writing on them: "These things don't really degrade in landfills, and of the 220 million tonnes of waste produced in the UK each year, about 25% is landfilled. Some is also shipped abroad."

Some say that the plastic guards currently in use should ideally be reused as many times as possible and eventually recycled or incinerated, with energy recovery. However, I know from experience that thousands of these guards are left out in the sticks and forgotten about for years. They essentially become litter. Collecting the guards after use is another expense in time, labour and money, which may pose a disincentive to some tree planters.

I asked Joe and James about the decision process that leads to using tree guards. They told me that because each site is very different, the decision to use them is not straightforward. Notably, many tree-planting sites (e.g., for urban development mitigation and conservation projects) likely use tree guards when they're not needed. The main reason for them is to protect young saplings from grazing herbivores such as deer, rabbits and voles: the young shoots are highly nutritious and likely to be less bitter and more tender. However, if there's no significant grazing pressure from these mammals, tree guards might not be needed. If local deer populations are relatively low but vole and rabbit populations are high, a deer fence around the site plus small biodegradable cardboard tree guards is one option. If the populations of all grazers are low, which is often the case, and other pressures such as wind and vandalism aren't an issue, the trees may grow just fine without protection. Joe told me that before considering using tree guards, a herbivore impact assessment is vital, which unfortunately doesn't happen enough. This involves surveying a site to determine which animals (e.g., deer, hares, voles, rabbits) live there and in what numbers.

I love deer, but their populations are huge in the UK. There are no natural predators (we've wiped them out), and it's often impractical to 'rewild' a site with predators because the landscape's so fragmented. Many sites are also close to urban areas, so even if wolf populations could survive in pockets of the countryside, it's unlikely public opinion would be on their side. This opens a debate on whether we should cull more deer. Personally, I don't eat meat, so my position is biased, but venison is more sustainable than other forms of meat – so there's an argument for more culling. Joe told me that the Woodland Trust has started to use drones with thermal imaging cameras attached to do herbivore impact assessments. This allows the rapid detection of deer and other mammals from the air, thus providing population numbers. He said, "With the drone, you can look at the screen in real-time and see blobs of light where the thermal imaging camera detects the heat from the deer, and you can zoom in to see exactly what species is present". He thinks they underestimate the deer population size from the ground by around 50%. In other words, with the accuracy of the drone, the population size is often double their estimate. This information helps them to determine which protection measures to install around the trees.

In terms of small mammals, many sites see raptors moving back into the habitat, which helps to maintain the equilibrium. Owls and kestrels enjoy the open grassland habitat surrounding the trees where voles and mice hide, so we should encourage raptors into these sites. To encourage them, it might be appropriate to install raptor nesting boxes on or near the site. Owing to habitat clearance, nest site availability can be a limiting factor for raptor population: many use hollows in veteran trees, but these are becoming increasingly scarce. In *Invisible Friends*, I wrote about how a regenerative agriculture farmer had an issue with moles. They tunnelled up to the surface and knocked out his courgette plants every night. Yet, because his philosophy was to promote biodiversity across the site, owls eventually moved into the local trees. One morning, he was walking his dog when the owls swooped down and grabbed the moles. This is part of the natural order of things, and the equilibrium in the ecosystem was restored.

Another thing Joe mentioned was that people often place tree guards around thorny shrubs – but they don't really need the guards: they've already evolved a solid defence strategy and sprout regrowth more easily than trees. However, I spoke to one person on a social media platform who said that deer and hares chomp right through the thorns and prefer to eat the regrowth, thereby killing the shrub. Using tree guards is such a contentious topic that people will likely have different experiences in different parts of the world. One person told me they simply stack deadwood (old sticks from trees) over the newly planted tree sapling.

So are tree guards necessary? In some instances, they enhance a tree's ability to survive in the early years. In other cases, fences or even no measures at all may be fine, depending on the herbivore grazing pressures. It seems that Joe's suggestion of a herbivore impact assessment is a vital first step when deciding if they're needed. One thing is certain: there's not a one-size-fits-all solution when it comes to protecting tree saplings. Are tree guards safely biodegradable in natural conditions? In other words, are any of them ecofriendly? As demonstrated by the Woodland Trust's trial site, there are many varieties of guard. As already noted, the plastic ones will take many years to biodegrade fully – and to do so, the plastic must first become microplastic, which can damage the ecosystem. But what about the so-called ecofriendly ones? Some prototype guards are made from thick cardboard. As you can imagine, these biodegrade well, but they're not as robust as the plastic alternatives in the field, particularly in wetter climates where they go soggy. Some guards are made from pulp, similar to the cardboard ones, but their chunkier composite nature appears to increase their robustness. They're relatively cheap, can be made from recycled paper and have integrated pulp hoops around a bamboo stake that holds the guard in the ground. This is a good feature because typical guards require plastic cable ties to hold them to the stakes, thereby increasing plastic waste.

Some of the new tree guards are made of cotton and pine oil. Although these are robust, it's not known how long they take to fully biodegrade: because they're so new, they haven't been tested in the field yet. There's also a question of the sustainability of the cotton material they include. According to Joe, it's not organic, and non-organic cotton is damaging to the environment.[7] It can take 2,700 litres of water to produce the cotton required to make a single T-shirt,[8] which has an negative impact on local ecosystems. Intensive cotton production can also lead to severe soil degradation. Some have said this makes it one of the "world's dirtiest crops".[9] Traditional cotton farming is widely cited as accounting for 24% of the world's insecticide sales and 11% of global pesticide consumption – just one crop![10] I'm less clear about the sustainability of the pine oil used in the tree guards. Yet the cotton figures alone trigger alarm bells.

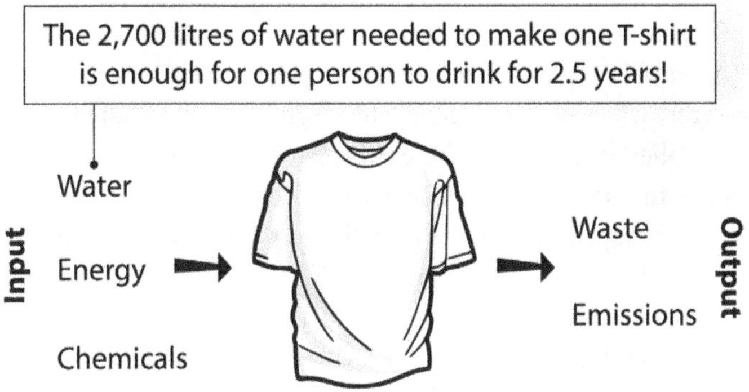

The environmental impact considerations of cotton.

There's also a new guard made from wool and cashew nut oil. These are robust in the field, although slightly brittle. This might be good because it suggests they'll eventually degrade quicker. However, there are human rights questions around cashew nut production. According to a report by Human Rights Watch, in some countries the cashew industry relies on forced labour and the exploitation of children to keep costs down.[11] In Vietnam, forced labour was documented among vulnerable community members, including people in prison and on drug charges, where the work is referred to as 'rehabilitation'.[12] I can't verify this, but according to several reports, if these people refused to work they were punished with torture or solitary confinement. These reports should be interpreted with caution, but they trigger alarm bells for cashew nut oil that's used in 'sustainable' tree guards. Moreover, the shelling of the nuts is often done by hand. This method's strenuous, and the acids required to break the hard shells cause painful hand injuries. Again, alarm bells are ringing, and this should all be investigated before endorsing cashew nut oil as a suitable alternative material for tree guards.

Finally, the cashew plant's a thirsty one, requiring lots of water. It can take more than 45,000 litres of water to produce a single kilogram of cashew nuts.[13] Like cotton production, this can put pressure on the local ecosystem by draining the water from rivers, lakes and underground sources. This process can increase the soil salinity and damage soil quality. Water-intensive crops also prevent local people from accessing the water that's needed for food security and public health. According to a United Nations report, the share of water accessible to each person has declined by more than 20% in recent decades.[14] This is another reason to be cautious when selecting replacement materials. We tend to find a solution only to create another problem.

There's also a question of what happens to the cashew oil (or other substances) when the guard finally biodegrades. Does it affect the soil? The microbes in the soil? The earthworms and other tiny creatures that keep the soil healthy? These are all questions that need to be asked before supporting the mass roll-out of a new tree guard. A final example at the West Country trial site is a guard made of a polylactic acid (PLA) composite and corn starch. These seem fairly robust, but ultimately the PLA is still a type of plastic – a bioplastic. However, it's typically made from renewable sources, so its sustainability rating is much better than 'normal' plastics. Despite being more sustainable, the Woodland Trust doesn't fully understand the effects of PLA on the ecosystem when it degrades. For example, will the chemical changes that occur upon degradation affect soil organisms? Moreover, several studies have shown that as PLA degrades it becomes a microplastic, and sometimes this exhibits the same effects as a conventional microplastic.[15] Further research is needed to see if bioplastics harm the ecosystem; once again, we may find that they solve one issue only to cause another. While researching tree guards, I asked several restoration ecologists if they had experience planting trees and whether they used guards. Some people told me they didn't use guards or even plant trees because 'natural regeneration' was sufficient. This is probably the most ecofriendly way of restoring forests – no plastic, no carbon footprint from tree nurseries and transportation, and no money spent. It's likely, though, that herbivore pressure was low to negligible in these instances. I also posted a Twitter post to ask for people's views on tree guards. Some of the comments were helpful, and they certainly highlighted the contentiousness of this topic.

There are many tree guard options, and we should always strive to use the most ecofriendly ones – and watch out for greenwashing. But the initial question should be, "Do I need to use them?"

Chewing Gum – I Once Was a Tree

Chewing gum isn't something I do nowadays, but I remember as a child buying many packets of the stuff and chewing incessantly. Ever wondered where the rubber in chewing gum comes from? Chicle, the traditional ingredient used to give gum its chewy texture, originates from the sapodilla tree. Scientifically known as *Manilkara zapota*, this is native to Central America. The tree's bark, when tapped, releases chicle, a sticky and milky sap. This is harvested by making incisions in the bark, allowing the chicle to flow out. Once collected, the sap is purified and processed to remove impurities, resulting in a natural rubber that forms the basis of chewing gum.

Chicle has been used for centuries in various applications, including as a chewing gum ingredient by ancient Mayan civilisations. Today, while synthetic alternatives are often used in commercial gum production, chicle still holds a special place in the hearts of gum enthusiasts who appreciate its natural origin and unique history.

PART IV

UNSEEN AND UNDER-APPRECIATED PERSPECTIVES

CHAPTER 13

Trees and Their Invisible 'Friends'

"Invisible threads are the strongest ties."
—Friedrich Nietzsche

Trees and other visible organisms are far from alone in a forest ecosystem. In fact, they're dwarfed by the sheer number of invisible life forms that dwell below, within and around them. The unseen biodiversity in the subterranean realm is vital to the forest's health. It's part of the forest. Indeed, a forest is as much invisible as it's visible.

When thinking about ecosystems, I like to remind myself of Louis Pasteur's quote: "The role of the infinitely small in nature is infinitely great."[1] Indeed, beneath the surface, in the depths of the soil and within the tissues of plants, trillions of microbes scurry to and fro, engaging in a myriad of biochemical interactions. They forge unseen networks of life that play roles in the forest's overall health and vitality. Coupled with an array of biogenic compounds, these tiny entities help shape the ecosystem, influencing everything from nutrient cycling to disease resistance. The beauty and complexity of a forest, therefore, is as much a product of this invisible world as it is of the visible one.

Nutrient cycling

In forests, like in all ecosystems, microbes are involved in cycling nutrients, making them available to the forest's trees, fungi, plants, and animals. Microbes supply nutrients to trees by converting nitrogen, sulphur and phosphorus and other compounds into plant-friendly forms.[2] They do this by degrading organic matter. This is anything that contains carbon, such as the remains of animals and plants – the leaves that float to the ground each autumn in deciduous woodlands, the twigs and bark that fall, the plants (and animals) that eventually wilt and die. Organic matter, being the primary source of energy for these microorganisms, fuels the biological activity in the soil, leading to nutrient cycling and enhanced soil fertility.[3]

Frankia and alder

Deciduous trees called alder in the genus *Alnus* grow in the temperate forests of Europe and America. They can reach over 25 m in height and typically live for around 60 years. The bark of the alder is dark and fissured. It's often covered in lichens and mosses, and supports many insects – beetles, butterflies and moths – including the adorably named alder kitten moth. Alder is monoecious, meaning both male and female flowers are found on the same tree (dioecious means individuals have either a male or female reproductive organ). In early spring, the alder trees put on a magical display as they sprout long, drooping catkins. These resemble delicate tassels, swaying and adding a unique textural element to the tree. As the seasons change, the catkins give way to tiny, cone-like fruits, each a potential new life, ready to continue the alder tree's legacy.

Its natural habitat is moist ground near rivers and lakes and cool, wet, marshy grasslands. Like other pioneering trees such as birch, alder can seed and quickly colonise bare ground. Alder trees also have what seems like a magic trick up their sleeve: they can rapidly colonise nutrient-poor soils where few other trees thrive.[3] This ability primarily occurs because of the tree's relationship with a symbiotic nitrogen-fixing bacterium called *Frankia alni*.[4]

This relationship occurs exclusively between *Frankia* bacteria and trees in the *Alnus* genus. *Frankia* 'infects' the alder's root hairs and starts forming nodules around the roots. The partnership has a beautiful give-and-take dynamic. The tree provides the *Frankia* with sugars, products of its photosynthesis, and in return, the bacterium accomplishes an extraordinary chemical feat to create the root nodules. It captures nitrogen from the air – a form unusable by plants – and converts it into a form that plants can readily absorb and use to build proteins. This process, known as nitrogen fixation, is akin to turning air into food. Research suggests that *Frankia* provides most of the tree's nitrogen requirements.[5] This mutually beneficial relationship (whether it's a 'friendly' one is up for debate) also improves the fertility of the soil, thus shaping the dynamics of the surrounding ecosystem. Imagine a symbiotic partnership so profound that it fundamentally alters the very environment in which it exists.

Evidence suggests that the Inca peoples in pre-Columbian South America recognised this relationship between *Frankia* and alder trees. They knew the tree increased soil fertility and stabilised terrace soils in their upland farming systems.[6]

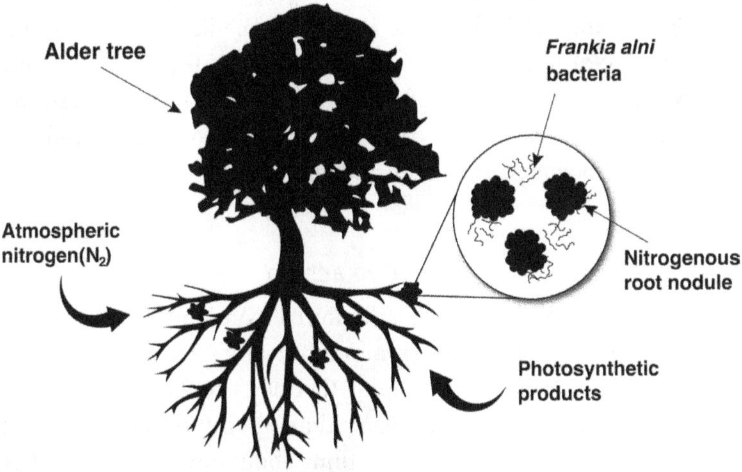

The mutually beneficial relationship between Frankia *and alder.*

We can think of microbes as the sustainable farmers of the forest ecosystem. They're also the recycling plant operatives, miners and supermarkets of the plant world, and forest ecosystems simply would not exist without them.

Microbes and forest immunity

Microbes also play a role in a forest's immunity and resistance to pathogens. Beneficial microbes living on tree leaves can help stave off infections by regulating a plant's immune response, outcompeting opportunist pathogens and producing antibiotic chemicals.[7] As in the human body, tree microbes act like tiny chemical factories, sending messages to express genes that drive pathogen resistance. Some bacteria can degrade the cell walls of pathogenic fungi and alter their proteins.[8] This has the effect of reducing fungal infections. Certain members of tree microbiomes can produce antibiotics and play a role in outcompeting pathogens. I often use the following analogy: Imagine one medieval army (the pathogens) storming the castle of another army (the protectors). They both want the same thing – the riches and rights of the kingdom – but the protector army is strong and unwilling to give up this prize, despite the forthcoming assault. A tree or plant microbiome with more species has been linked to improving the plant's capacity to suppress diseases. An increase in species diversity often (but not always) corresponds to more ecological functions being fulfilled. Think about our medieval armies again. If our protector army has a greater variety of soldiers and weapons, it may have a better chance of defending the castle. In contrast, the castle may fall if its defences can only draw on foot soldiers and not archers, crossbowmen, spearmen and gunners.

Collective action

The leaves and bark of trees host a diverse community of microbes known as the phyllosphere (pronounced 'fillo-sphere').[9] Some of these microbes produce compounds that deter herbivores and limit the growth of pathogenic organisms, thus serving as another line of defence for the tree. The fascinating thing about these microbial allies is that they don't just benefit individual trees: through their collective action, they contribute to the resilience and health of the entire forest ecosystem. Whether this is driven by 'group selection' is a whole different argument. This refers to a proposed mechanism of evolution in which natural selection is imagined to act at the level of the group rather than at the more conventional level of the individual or the gene. In this hypothesis, a trait might be favoured not because it benefits the individual organism or its close relatives but because it benefits the survival or reproduction of the group as a whole (in our case, the forest rather than the individual tree). This could, in theory, lead to the evolution of advantageous traits at the group level, even if they are disadvantageous at the individual level.[10] However, the concept of group selection is controversial among evolutionary biologists.[11] While it has been used to explain certain behaviours, especially in social animals, many scientists argue that other mechanisms, such as kin selection or individual selection, can better explain most observed phenomena.

Whether or not forest immunity is selected for at group level or individual/gene level, we must recognise the invaluable role of these microscopic defenders when we consider the immunity of forests. Without them, our forests wouldn't be the often vibrant and humming communities they are.

The Future Oak Project

Future Oak is a pioneering project started in the UK in 2021. It aims to investigate the role of beneficial microbes in fighting diseases that affect Britain's native oak trees.[12]

Quercus robur, known as the English or pedunculate oak, is a species of flowering plant in the family Fagaceae. It's native to most of Europe, extending eastwards to the Caucasus and the Caspian Sea. A slow grower, it takes time to mature, often centuries, but once it does it's a sight to behold, and is often touted as the British people's favourite tree. Picture a large, majestic tree, its strong trunk expanding several metres in diameter, while its broad, twisting branches provide shelter and food for a myriad of creatures. Indeed, according to the Woodland Trust, native oaks – including *Q. robur* and sessile oak (*Quercus petraea*) – can support hundreds of different invertebrate species. One of the most widely accepted estimates is from a study by Kennedy and Southwood (1984), which reported that Britain's native oak species support around 284 different species of insect herbivores alone.[13] This figure doesn't even include other invertebrates such as spiders, mites, and woodlice, among others. The actual number may be even higher, considering factors such as geographic range, local environmental conditions and interactions with other species. And remember that oak trees, like other trees, support billions of microbes too.

Oak trees create a complex ecosystem within and around themselves, serving as a vital hub for biodiversity. They provide food (and habitat) for invertebrates, serving as a food source for various birds and mammals. Each of these invertebrates and microbes plays a role in the ecosystem, contributing to processes such as nutrient cycling, pollination and serving as a food source for other wildlife. Therefore, in one sense, the oak tree is not just a tree – it's an ecosystem engineer. Its gnarly bark, rough and furrowed, is also home to various mosses and lichens, hinting at the nested symbiotic relationships that occur in nature.

The English oak is deciduous. Its vibrant green leaves are broad and lobed, creating a breath-taking canopy during the summer months, only to burst into a fiery spectacle of yellows and browns during autumn. When I described the English oak to the students in a Plant Ecology lecture at Flinders University, South Australia, I referred to its 'peduncle' as a good ID feature. Peduncle is a word that invariably receives a few chuckles but simply refers to the stalk on the acorns. The sessile oak (also native to Britain), on the other hand, has its acorns attached directly to the branch without a peduncle, hence the term 'sessile', which means 'sitting' in Latin.

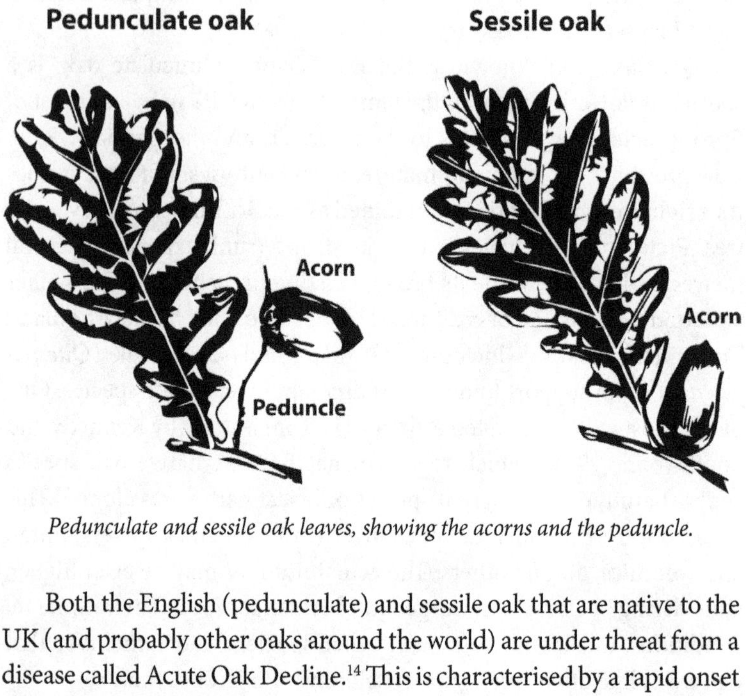

Pedunculate and sessile oak leaves, showing the acorns and the peduncle.

Both the English (pedunculate) and sessile oak that are native to the UK (and probably other oaks around the world) are under threat from a disease called Acute Oak Decline.[14] This is characterised by a rapid onset of symptoms, typically affecting mature trees over 50 years old. One of the most notable of these is a dark fluid that oozes from cracks in the bark and trickles down the trunk, often accompanied by crown thinning and dieback. The cause of Acute Oak Decline is not entirely clear, but it's thought to involve a complex interaction of different factors. A bacterial infection, particularly from species of the *Brenneria* and *Gibbsiella* genera, appears to be involved, possibly in conjunction with environmental stressors. It's thought that climate change also contributes to the onset of the disease: trees in warmer and dryer climates are probably more likely to suffer from Acute Oak Decline because extreme conditions such as drought events (which are increasing because of climate change) weaken the trees to the point that they cannot fight off the pathogens. Although the biochemical mechanisms are different, it's a bit like when humans endure prolonged stress, which can suppress their immune system and promote inflammation. Over time, this immune suppression can make a person more susceptible to infections, and wounds take longer to heal. Conservation efforts to combat Acute Oak Decline involve ongoing research to better understand the disease, its causes and potential treatments. This is where the Future Oak project comes into play. Researchers in this project team recognise that, much like in humans and other organisms, oak trees have a microbiome that strongly benefits their health. The diverse collection of microbes living in and around the trees, the ways they interact with each other and the conditions in which they occur define the tree's microbiome. The Future Oak project's core aim is to better understand the identity of these microbes, what they do and their role within the oak microbiome and immunity. The team uses state-of-the-art DNA sequencing technology to identify the microbes growing on oak trees. Once they identify those that can boost the tree's immune system, they'll grow them in the lab. They'll then trial the disease-suppressing microbes and test their ability to prevent the progression of Acute Oak Decline. To do this safely, they'll infect oak logs and saplings with Acute Oak Decline in controlled lab conditions to prevent the pathogen from escaping, and then inoculate the infected logs and saplings with potentially beneficial microbes and hope they can fight the disease.

Here's to hoping the forest's invisible friends (and their visible counterparts – the researchers) can help stave off Acute Oak Decline in the face of mounting environmental pressures.

Talking trees and the underground marketplace

In my first book, *Invisible Friends*, I wrote about the research of forest biologist Suzanne Simard. She studies the potential communication and cooperation between trees in a forest community through what's often called the 'wood wide web' – a proposed vast network of fungal mycelium connecting trees.[15] Simard's work suggests that trees in a forest communicate with each other in mutualistic ways through this underground biological network. Her work counters the prevailing notions of pure competition that have long dominated ecological sciences. The historical perspective has painted a picture of different species locked in a relentless battle for light, water and nutrients; but the pioneering work of renowned and late evolutionary biologist Lynn Margulis, who tirelessly advocated for the power of cooperation in species evolution, is reinforced by Simard's findings.[16]

According to Simard, this intricate underground network allows a forest to function as a unified organism, harmoniously intertwined. In this narrative, the trees communicate using a language of gases, hormones, vibrations and water, tapping into an immense reservoir of information to support and nurture each other. Simard reflects on the inherent cognitive bias of human beings, who forever place themselves at the pinnacle of a hierarchical structure and consider themselves the superior species. This bias closes our eyes to the complexities and interrelated ecologies of other life forms. To find it astounding that organisms possess different forms of intelligence and engage in communication and cooperation is to succumb to this cognitive trap. Instead, Simard invites us to appreciate the beauty of this intricate web of life, acknowledging the remarkable abilities and interconnectedness of all living beings.

Although the notion of conviviality and resource 'sharing' between trees and their underground partners has been questioned, there's no doubt that the beauty and complexity of forest ecology will remain unblemished.

The wood wide web

Scientists have, for a while, believed that the subtle exchange of information between trees is facilitated by near-invisible mycorrhizal fungi. These fungi, existing as delicate threads known as hyphae, interconnect to form a sprawling network called mycelium. In a thriving forest, this mycelial web can be so intricate and dense that it spans hundreds of kilometres beneath even a single footstep. Some scientists think these mycorrhizal networks not only connect trees of the same species but also forge connections between different species, such as the birch and fir. This expansive and bidirectional flow of information has drawn comparisons to the vastness of the internet, leading to the catchy term 'wood wide web'.[17]

Mycorrhizal fungi are able to establish connections between plants in various ways. One such pathway involves ectomycorrhizae, which encompass the external surface of tree roots (hence 'ecto', meaning 'outside').[18] Another pathway employs endomycorrhizae, which are able to penetrate the interior of tree roots (hence 'endo', meaning 'inside'), intertwining their hyphae between the cell wall and cell membrane – almost like wedging something between a bike tyre and an inner tube.[19] Arbuscular mycorrhizae, yet another type of mutualistic fungi, reside within tree roots and extend their hyphae to form expansive networks, facilitating nutrient uptake from the soil. According to Simard, there are vital players in this intricate network known as hub trees or, more affectionately, mother trees.[20] These remarkable entities, Simard maintains, nurture the young saplings that populate the forest's understorey. Hub trees establish connections with many other trees in the forest through the dense network of microorganisms. They act as conduits, transporting excess carbon to the seedlings and significantly increasing their chances of survival. Trees are also thought to communicate warnings about stressors and bolster each other's resistance to pathogens through this intricate network. Beyond mycorrhizal fungi, another group of cooperative microbes, known as mycorrhiza helper bacteria, plays a role in tree health and communication.[21] These diverse bacterial groups can enhance mycorrhizal function, stimulate growth, facilitate nutrient uptake, improve soil conductivity and boost resistance against pathogens. In a fascinating study, these helper bacteria were found to assist fungi in forming mutualistic relationships with trees.[22] It's hypothesised that these bacteria aid trees and other plants in defending themselves against pathogens by enhancing nutrient uptake and enabling plants to allocate more resources to their defence mechanisms. Thus, the immense networks of microscopic soil fungi not only facilitate communication, growth and defence mechanisms in trees but also receive assistance from an entirely different kingdom – the bacteria. It's a spectacular arrangement that has evolved over thousands of years, forging cooperation between the micro- and macro-worlds. Or so the popular 'cooperation' narrative goes – a narrative that I once found compelling. But a growing volume of voices has raised concerns about that narrative, and I for one welcome this balance.

Critique of the 'sharing and caring' narrative

First, one thing that's unquestionable is that as we walk through a woodland there's a dynamic spectacle beneath our feet, obscured from our eyes: a world tucked within our world, an interwoven matrix of infinitesimal life forms communicating through a dense system of

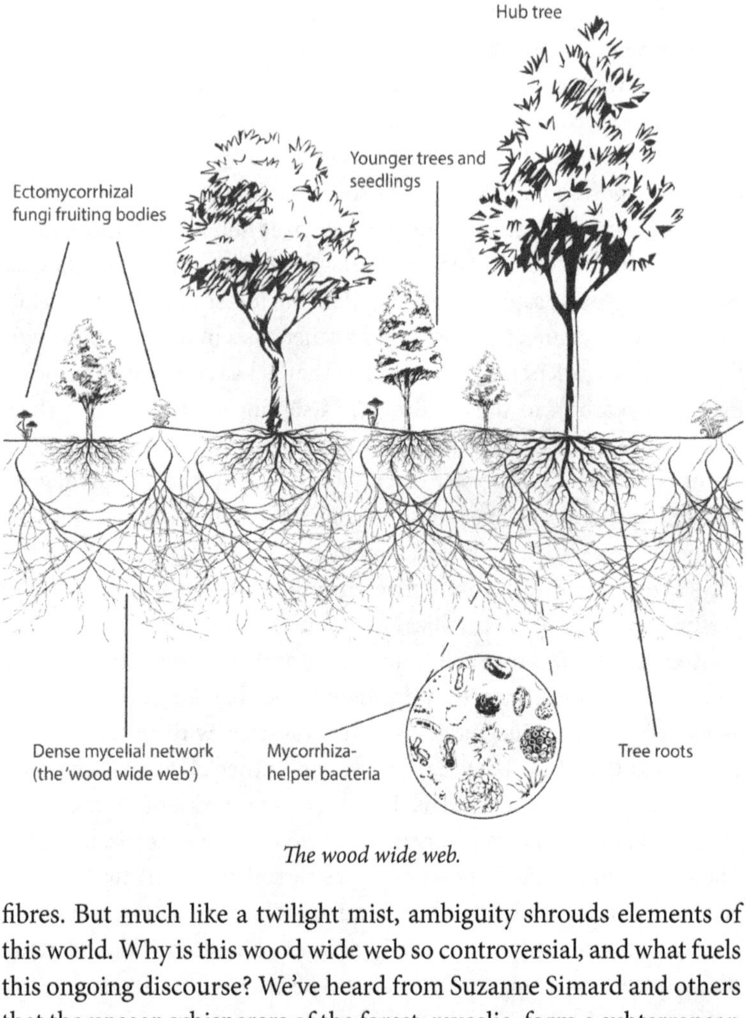

The wood wide web.

fibres. But much like a twilight mist, ambiguity shrouds elements of this world. Why is this wood wide web so controversial, and what fuels this ongoing discourse? We've heard from Suzanne Simard and others that the unseen whisperers of the forest, mycelia, form a subterranean network that links the roots of trees and plants, facilitating the trade of nutrients and information, a mutual symbiosis on an almost unfathomable scale. In this line of thinking, the complex network gives voice to a silent discourse beneath the forest floor. Yet the realm of science is constantly under the sway of 'why'. And in this case, the 'why' pertains to the true nature of this subterranean network.

Critics argue that the wood wide web (if it exists) is not solely a philanthropic endeavour. Some say there's nothing 'friendly' about it. Evidence suggests instances of certain dominant trees, the proverbial elders of the forest, siphoning off more resources. Sceptics contend that the wood wide web concept may have spun a web of oversimplification.[23] Forests, some maintain, are battlegrounds of survival, where competition reigns supreme.

Toby Kiers

Evolutionary biologist Toby Kiers based at VU University Amsterdam is one leading scientist who questions the mutualistic wood wide web narrative. Kiers has been involved in some innovative studies exploring the relationship between plants and subterranean fungi.[24] Her team has provided evidence that fungi act as powerful traders: they shrewdly amass nutrients, turning their subterranean world into a treasure trove. Plants with a generous bent, sharing their carbon wealth freely, find favour in the eyes of these microscopic power players, earning them rewards. Conversely, those plants that are tight-fisted with their carbon resources face fungi's subtle wrath, a gentle nudge towards generosity.[25] But the fungi's masterstroke lies in their game of resource management. They navigate the currency of nutrients, moving and trading with deft precision to secure the most advantageous 'deal' for themselves. In this narrative, it's a world of trade and barter where carbon and nutrients change hands, a world dictated by the laws of give and take and not necessarily the sharing and caring kind. Early in her career, Kiers was inspired by a famous 1968 essay in the journal *Science* called 'The Tragedy of the Commons' by ecologist Garrett Hardin.[26] This explored a dilemma arising from the situation in which multiple individuals, acting independently and rationally according to each one's self-interest, ultimately deplete a shared limited resource, even when it's clear that it's not in anyone's long-term interest for this to happen.

Hardin uses the metaphor of a common grazing pasture, open to all who own livestock. Each herdsperson, acting in their self-interest, will want to put as many of their own animals on the commons as possible. While each additional animal a herdsperson adds yields a benefit to the individual, it also takes a toll on the shared resource, depleting it a bit more. While the benefit of an additional animal is fully received by the herdsperson, the damage to the commons is shared among many. Consequently, each herdsperson, in their self-interest, will keep adding livestock, leading to overgrazing and destruction of the commons. This is the tragedy: each person pursuing their individual benefit leads to an outcome that's detrimental to all.

Hardin's essay is often interpreted as a call to adopt global policies to manage shared resources, and it has had a significant influence on thinking about conservation, population growth, resource management and the concept of sustainability. It has also inspired various scientific experiments. Kiers and her team studied legume plants and their interactions with nitrogen-fixing bacteria.[27] They knew that legumes host many types of these nitrogen-fixing bacteria, and any one of these bacterial strains could fix nitrogen while exploiting the sugary exudates of the plants. They started wondering if some bacteria 'cheated' by letting the others fix the nitrogen for the plants while still benefiting from the plant's sweet nutrients. They ran computer models to show that if the bacteria did indeed 'cheat' their plant hosts (the legumes – which can be viewed as a 'commons' in this case), then this supposed mutualistic relationship would fall apart as more and more bacterial strains cheated.

The team ran some other intriguing experiments. They prevented a normally mutualistic strain of bacteria from cooperating (i.e., fixing nitrogen) with its plant host by replacing the air supply around the roots with a nitrogen-free form of air. They viewed this forced non-cooperation as analogous to cheating on the commons. Their results were exciting. They found that the plant reacted to this simulated cheating by shutting down the oxygen supply to the bacteria, significantly reducing their reproductive capacity.[28] The team referred to this as 'sanctioning'. Their results support the hypothesis that legumes select for more cooperative bacteria by imposing sanctions based on the amount of nitrogen that the bacteria fix. This shows that plants act in their own self-interest, as expending energy to supply bacteria with nutrients when the bacteria are not reciprocating is not beneficial to them.

But what about fungi? Kiers and her team also ran some controlled experiments on nutrient-sharing between plants and fungi.[29] Intrigued by the rules of this symbiosis, they tweaked the levels of cooperation in the plants and fungi, leading to a fascinating discovery. The plants could identify, differentiate and even bestow more carbohydrates on the most efficient fungal partners. In a similar vein, the fungal comrades repaid the favour by channelling more nutrients towards the roots that supplied more carbohydrates. Their results show that this symbiotic relationship is not the master–slave dynamic seen in some other mutualisms. Instead, it exhibits an evolutionary stability that hinges on the mutual control and reward of both parties. Thus, the partners who provide the most favourable exchange rate receive the most benefit – a beautifully balanced reflection of the equilibrium of nature. Kiers started constructing an economic model to better understand the interactions between plants and fungi. She proposed a theory by drawing insights from the mechanisms of the free market system. She suggests that the secret to the resilient plant–fungi mutualisms, which have thrived for an impressive 470 million years, does not lie in individual organisms devotedly acting for the communal benefit. Instead, the key seems to be mutual self-interest. In most scenarios, both plants and fungi reap greater advantages by exchanging resources rather than hoarding them. In essence, trade proves to be more beneficial than isolation, a concept that echoes in the rhythms of nature as much as in the world of economics. The team also found that mutualistic relationships between plants and fungi can break down if the host finds other ways to acquire its nutrients, for instance, by turning parasitic and feeding off other plants, or even carnivorous, feeding on insects.

Anthropomorphism and altruism

It's important to say that a relationship can be mutualistically beneficial without being 'friendly' in the typical way that humans see it. Critics of the communitarian view of the forest have warned against the seductive allure of anthropomorphic language (attributing human qualities to non-human entities). Do trees engage in elaborate conversations, or are we projecting our human yearnings onto their tangled roots? Critics argue that romanticised portrayals of this concept could lead astray those seeking to conserve and manage forests. With a flick of the pen, forests might be painted as harmonious utopias, neglecting the tangled web of complexities they truly are. There's a fear that oversimplification might cast a dark shadow over the nuances of forest dynamics, hindering a comprehensive understanding and misguided decision-making.

It's amusing, though, when journalist or scientist critics say the language of those who support the communitarian view of the 'wood wide web' is too anthropomorphic. I've noticed that in their counter-arguments they talk about fungi as 'market traders' and 'punishers' who 'reward' plants that are 'generous'. All these terms are typically attributed to human behaviours.

Anthropomorphism therefore clearly occurs on both sides of the argument. Furthermore, the notion that a non-animal organism cannot be 'friendly' with another because its behaviour is not altruistic requires more thought. After all, there's a strong argument that even our human 'friends' are not truly altruistic. Altruism refers to a selfless concern for the wellbeing of others. It's when one organism acts out of a desire to help others without expecting anything in return; or, in evolutionary terms, it's when a behaviour reduces the fitness of the actor but increases the fitness of the recipient. The question of whether human beings, including our friends, can be truly altruistic is a complex one and touches on aspects of psychology, philosophy, sociology and evolutionary biology. The question of whether plants and fungi can be altruistic is also steeped in complexities.

From an evolutionary standpoint, some scientists argue that what appears to be altruism can often be explained by concepts such as kin selection (an organism helps others who share their genes, such as relatives) and reciprocal altruism (an organism helps others in the expectation that they will help us in return when needed).[30] In the context of friendship, actions that seem altruistic may often be influenced by a kind of reciprocal altruism. Human friends typically develop mutual bonds and tend to support each other over time. So, while a friend might help you out without expecting immediate payback, there's often an underlying understanding that you would do the same for them if roles were reversed. This doesn't diminish the value or sincerity of their actions, but it does suggest that pure altruism might be rare. Hence, cooperating and, on some level, 'expecting' reciprocity could still be considered a 'friendly' interaction.

Plants, fungi and animals are, to some degree, personified in some cultures. This personification can take many forms and apply to animals, plants, rivers, mountains and other elements of the natural world. For instance, in many Native American cultures, non-human animals are often seen as spiritual beings or helpers with their own personalities and wisdom to impart. The concept of 'animal spirits' or 'spirit animals' is a common theme. Similarly, many cultures also see the Earth itself as a living entity, often referred to as Mother Earth, who provides sustenance and life. In Māori culture in Aotearoa (New Zealand), there's a deep connection with the environment and a belief in the spiritual personhood of natural elements.[31] This has even led to legal recognition of this status in some instances, such as the Whanganui River and Te Urewera, a former national park, which have both been granted the same legal rights as human beings. Whether or not this way of thinking is globally supported, it does show that we should be careful with how we frame and discuss anthropomorphism.

Back to Kiers

Kiers is a critic of the cooperative view of trees and microbes, and she prefers to focus on quantifying the trade strategies between them. However, she recognises that fungi have the potential to form a vast and complex underground network that connects trees, their roots and their surrounding microbes – called a common mycorrhizal network or a CMN (the thing many people call the wood wide web). Kiers has referred to this network as a massive underground circulatory system. When I caught up with Kiers, she told me that she never uses the term wood wide web for two reasons. The first is that it makes fungi sound like passive cables, accessories of plants. The second is that mycorrhizal networks are important far beyond woods: they are found in nearly all terrestrial ecosystems and are very important in wetlands, grasslands and elsewhere. While current technologies are not advanced enough to follow connections of individual threads of mycelium between trees in a forest, Kiers' experiments allow her to study trade at precise levels in the lab. Indeed, she and her team have even demonstrated the existence of sophisticated trading strategies by attaching fluorescent nanoparticles to nutrients.[32] These were inoculated into a plant–fungi network and observed using a special microscope. They found that the fungi shuttled nutrients (phosphorus) from the patches of high resource availability to low resource availability, resulting in faster fungal growth in the nutrient-poor area. The reasoning behind this, Kiers' team speculates, lies in the concept of market forces at play in this microscopic world. In regions where phosphorus was scarce, the fungi could demand a higher 'price' from the plants, paid in the form of carbon-rich sugars. However, Kiers clarifies that their ability to trace the carbon's journey directly was somewhat limited.

Another curious finding in this study was that the journey of phosphorus wasn't a one-way street, merely flowing from the nutrient-rich area to the less endowed one. Instead, Kiers and her team observed an intriguing spectacle. The nutrients showed complex flow dynamics within the network. In particular, resource flows seemed to oscillate with remarkable regularity, happening every five minutes, forming a mysterious rhythm.

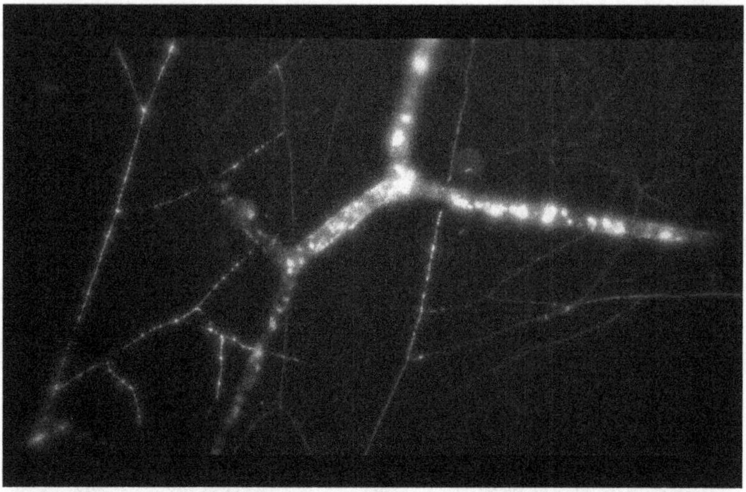

Flows of fluorescently labelled carbon in a mycelial network. (Image provided by Toby Kiers.)

Kiers' lab focuses on resource flows, not communication per se, whereas others have studied chemical signalling. The evidence for transmitting information (in chemical form) across these fungal networks underground is already strong. For instance, a 2013 study led by David Johnson at the University of Aberdeen demonstrated that bean plants (*Vicia faba*) infested with aphids (*Aphis fabae*) were able to send chemical warning signals to nearby plants via a network of symbiotic soil fungi, arbuscular mycorrhizal fungi.[33] The researchers set up several 'mesocosms' (a controlled experimental system replicating real-world conditions but on a smaller, manageable scale), each containing five bean plants. The plants in each mesocosm were grown in a shared pot of soil, enabling the development of a network of arbuscular mycorrhizal fungi linking the plants. To test their hypothesis, the scientists unleashed aphids on one plant in each mesocosm.

They found that the plants that were not under attack but connected to the infested plant through the fungal network began to produce defensive chemicals. These chemicals seemingly made their leaves less appetising to the aphids. However, in mesocosms where the fungal network was severed, the healthy plants didn't react to the aphid attack on their neighbour, indicating that the signal was sent via the fungal network. This study, along with Simard's work, showed that plants connect with each other. They seem to use the fungal networks in the soil to communicate and warn each other of potential danger. This helps plants to pre-emptively activate their defensive mechanisms, potentially improving their chance of survival when under attack. But what is the benefit to a given plant of sending a signal to its neighbour? Is it group selection? And what prevents partner plants from hijacking or manipulating information to their own benefit? These are the kinds of questions Kiers is working on: "It would be useful to have an evolutionary explanation for why sending chemical warnings would be in their benefit." So we know that underground fungal networks act as conduits of chemical signals shuttled between plants. But we don't know if these fungal networks can process the chemical information handed to them by the plants. More experimental work is needed to unravel the complexities of these underground interactions. Still, one thing is for sure: fungi are vital for forest health, whether they are being cooperative, spiteful, altruistic, reprimanding – or all or none of the above.

A future of underground discoveries

Kiers is now on a mission to map the world's underground fungal networks. She founded the organisation SPUN – the Society for the Protection of Underground Networks – to help her do so.[34] SPUN aims to "protect and harness the mycorrhizal networks that regulate the Earth's climate and ecosystems". They even provide 'Underground Explorer' funding to help people across the globe to map their local fungal networks.

I think the controversy surrounding the wood wide web is healthy. It signifies scientific discourse where minds clash, ideas evolve and knowledge expands. The intricacy of scepticism and analysis fuels the quest for a deeper understanding of the natural world. Scientists continue to venture into the depths of the forest, peering beneath the surface, collecting evidence and challenging assumptions. On the contrary, proponents claim that these very interactions balance the ecosystem, providing much-needed support to struggling trees and thereby maintaining the biodiversity of the woods. But despite the discordant notes in this woodland symphony, what remains undisputed is the integral role that this invisible realm plays in the beauty, complexity and survival of the visible entities.

CHAPTER 14

Senses and Memory

"Breaching out of their rooted darkness, breakthroughs are seeds of nonconformity determined to sprout into the light of free thinking."
—*Monica Gagliano*

So many brains, I think to myself. I'm walking around UCL's Grant Museum of Zoology and there's an entire section devoted to brains in jars. Each is preserved in alcohol and suspended by a thread. There's a pigeon and a turtle brain, but most are from mammals, collected from Africa, Asia, South America, Australia and Europe. Being surrounded by so many brains got me thinking about nervous systems, senses and cognition in our leafy friends. Do these phenomena exist?

Trees don't possess brains or central nervous systems, which we typically associate with the processes of sensing and remembering in animals. However, research implies that they can interact with their environment in complex ways that may be viewed as sensing. They are also able to respond to environmental changes in ways that might be seen as analogous to memory.[1] 'Sensing' in biology generally refers to the ability of an organism to detect and respond to certain stimuli in their environment.[2] These stimuli can be external, such as light, sound or temperature changes, or internal, such as biochemical changes. I would therefore argue that trees certainly are capable of sensing.

Trees can detect and respond to a range of environmental cues. For instance, they can sense light, which directs their growth, detect gravity, which helps them orient their roots downward, and respond to sound and touch (e.g., by growing thicker or changing direction when faced with physical obstacles).[3] Evidence suggests they can also sense chemical signals, such as those released by pests, and respond accordingly by producing defensive chemicals.[4] As for memory, it's important to clarify what this means. In human terms, memory usually implies a conscious recollection of past events. Evidence suggests that trees probably don't have this capability but can alter their growth patterns and physiological responses based on past experiences.[5] For example, trees can acclimatise to changes in climate. If a tree experiences a few years of drought, it might 'prepare' itself for future droughts by altering its water use strategies, growth rate or other aspects of its physiology. In a sense, the tree is using information from its past to inform its future responses – a process that might be considered a form of memory. We know that trees retain a record of history. For instance, it's widely known that their growth rings provide a historical record of environmental events. But are they capable of 'remembering'? Cutting-edge epigenetic research shows us that there's far more to this memory narrative than one might initially think.[6] We'll return to this later in the chapter. A tree's way of interacting with the world fundamentally differs from ours. Nevertheless, these differences do not diminish the importance or complexity of their responses to the environment. Many of the 'things' you use in life were 'once trees'. But did these entities also once have senses and memories?

The Umwelt

As mentioned in the Introduction, the concept of 'Umwelt' was first introduced by the Estonian biologist Jakob von Uexküll in the early twentieth century.[7] The term is German for 'environment' or 'surroundings', but in the context of biology and philosophy it has a more specific connotation. Umwelt refers to the perceptual world in which an organism exists. It describes how every organism, depending on its specific sensory and cognitive capacities, perceives and interacts with its environment in a unique way.[8] In other words, each organism has its own 'subjective universe' tailored to its specific needs and survival strategies.

For example, a bee's Umwelt includes ultraviolet light, which is used to locate flowers and nectar. This is a part of the environment that humans, who can't see ultraviolet light, don't perceive. Similarly, a dog's Umwelt includes a rich palette of smells largely imperceptible to humans. A bat's Umwelt includes high-frequency echolocation to navigate and find food.[9] This involves producing sounds and then listening for the echo of these sounds bouncing off objects in their environment. By doing this, bats can build a detailed mental map of their surroundings, even in complete darkness. However, these sounds are primarily in the ultrasonic range, which is beyond human hearing. Most bat species produce echolocation calls that range from 20 kHz to 200 kHz, while the hearing range of a healthy young human tends to be from 20 Hz to 20 kHz.[10] Bats' ultrasonic calls can be extremely loud, sometimes reaching up to 140 decibels, similar to the noise level of a jet engine at take-off![11] However, since these sounds are ultrasonic, we don't hear them. Bats have evolved 'noise-cancelling' mechanisms (via the middle ear muscle and SL1 gene) that protects their hearing:[12] they can essentially 'turn off' their hearing momentarily while emitting a loud call so they don't deafen themselves.[13] Therefore, while bats might seem silent to us, their world is filled with the echoes of their loud ultrasonic calls. This is a prime example of the concept of Umwelt – the unique perceptual world of each species. In the bat's Umwelt, echolocation plays a critical role, while it's an aspect of the environment that humans are largely oblivious to.

Meanwhile, birds have a unique way of experiencing the world, shaped by their specific biological and sensory capabilities. One of the most striking examples is how many bird species perceive the Earth's magnetic field, a sense known as magnetoreception.[14] This is crucial for their ability to migrate across long distances with extraordinary accuracy. Humans, lacking this sense, cannot do this without the aid of technology, making it an aspect of the bird's Umwelt that's entirely foreign to our bodies. Birds and other non-human animals also have highly developed vision, often seeing a broader spectrum of colours than humans. Some birds, such as raptors, have acute visual acuity, allowing them to spot prey from great distances. These capabilities are an important part of a bird's Umwelt and essential for its survival and navigation. Birds also use their hearing to navigate: they rely on sound to communicate, find food and be alert to predators. Certain birds, such as the Great Grey Owl, can locate prey under the snow by sound alone.[15] The Umwelt of birds is a symphony of magnetic pulls, a panoply of vibrant colours beyond human perception and a vast soundscape filled with information. Each element forms a unique part of their subjective universe, guiding them through their aerial journeys.

The concept of Umwelt emphasises the absence of a single 'objective' perception of the world. Instead, each species experiences its own version of reality, shaped by evolutionary pressures to prioritise certain types of information over others. This understanding challenges us to consider the complexities and subjectivities inherent in our interactions with the environment and other living beings. It's amazing to think how we're bombarded with constant streams of information in the form of sounds, light and chemicals that are utterly imperceptible to us. Uexküll wrote:

> Each house has a number of windows, which open onto a garden: a light window, a sound window, an olfactory window, a-taste window, and a great number of tactile windows. Depending on the manner in which these windows are built, the garden changes as it is seen from the house. By no means does it appear as a section of a larger world. Rather, it is the only world that belongs to the house – its environment.[16]

The Umwelt applies to all living organisms, including trees. Each tree species interacts with its environment uniquely based on its specific biological and physiological capabilities. These unique ways of interacting with and perceiving the environment constitute the tree's Umwelt. For instance, trees don't have eyes or ears as animals do, but they have other ways of sensing their environment. They can sense light using photoreceptor proteins. They also sense touch, such as the pressure of wind, and can respond by changing the direction of their growth (also known as thigmotropism or phototropism when light is concerned).[17] Moreover, trees and other plants can detect chemicals in the soil and the air. They are thought to communicate with each other and other organisms through chemical signals and even sound.[18] For example, when a plant is attacked by insects, it may release chemicals that signal other plants to ramp up their production of defensive compounds.[19] It's also thought that trees and other plants can detect the chemical signals from nearby plants and adjust their root growth to avoid competition for resources – although we need more research to confirm this. Trees also sense seasonal changes, such as variations in daylight length and temperature shifts, to guide their yearly growth cycles, flowering and seed production. Trees do this using special photoreceptors and an array of hormone signalling.[20] So, while a tree's

Umwelt might be quite different from ours or a bird's or a bat's, it's just as complex and suited to the tree's needs. Like all living organisms, trees have evolved to perceive and respond to their environment in ways that increase their chances of survival and reproduction. This underscores the significance of the Umwelt concept and the intricate and varied ways that life on Earth experiences the world.

Tree rings

Tree rings (also called annual rings or growth rings) are fascinating features that tell us a lot about a tree's life and the environmental conditions it experiences. Each ring represents a year in the tree's life. This new growth is visible when you look at a cross-section of a trunk. The ring usually consists of a light part and a dark part: the light part represents the fast growth in spring and early summer, while the dark part represents the slower growth in late summer and autumn. Tree rings can tell us about climate history.[21] Wide rings typically indicate years with plenty of water and good growing conditions, while narrow rings can indicate years with less favourable conditions, such as drought. This field of study is known as dendroclimatology, and the broader field of studying tree rings to determine historical events is known as dendrochronology. The term comes from the Greek words 'dendron' (tree), 'chronos' (time), and 'logos' (study), effectively meaning 'the study of time through trees'.

Because each ring represents a year, counting them can give a good estimate of a tree's age. In addition, they can be used to date events in human history, such as building construction. Moreover, trees living in the same area and therefore experiencing the same environmental conditions typically have similar ring patterns. Scientists can use these patterns to match up tree rings from different trees or even wooden artefacts, helping to provide a continuous record stretching back thousands of years. Tree rings can show evidence of forest fires and other events, such as volcanic eruptions.[22] For instance, a fire might cause a scar visible in the tree's rings, while a volcanic eruption could lead to a narrow ring owing to the decrease in light reaching the Earth's surface. Tree rings can also indicate a tree's orientation. In some cases, tree rings can be wider on one side of the tree than the other. This is often caused by a tree leaning, which can cause more growth on one side as the tree works to strengthen and balance itself.

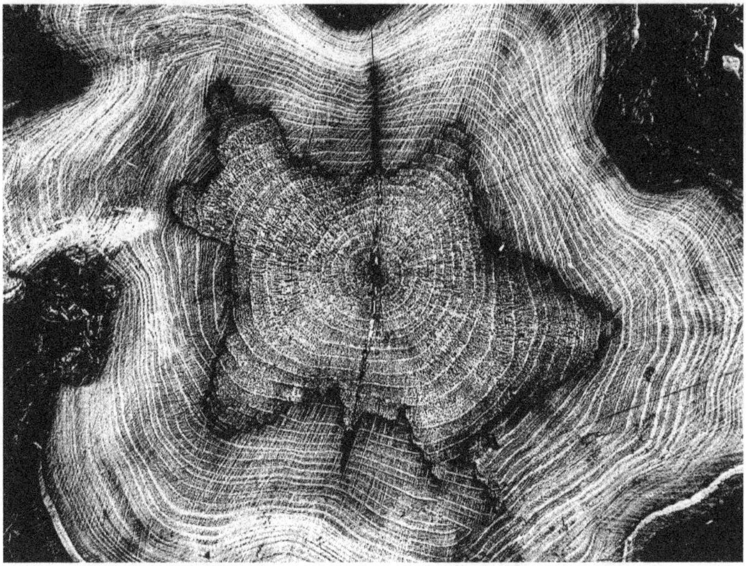

Tree rings can tell us important things about historical events.

A tree therefore keeps a diary of its development and the events that it lives through. Some events leave a bigger mark than others. For instance, most trees in the UK planted or naturally grown more than 50 years ago will have a tree ring diary entry that shows the extreme drought event of 1976, one of the most severe on record in the country.[23] This exceptionally dry period extended over 16 months, from the start of 1975 to the summer of 1976. It significantly impacted both people and the environment, making it a notable event in the UK's climatic history. The very high temperatures and extraordinarily low rainfall meant that the summer was one of the hottest and driest on record, leading to water shortages, restrictions on water use and impacts on wildlife and agriculture. The drought was so severe that the UK government appointed a Minister for Drought, Denis Howell, to manage the crisis, imposed a hosepipe ban and ran public information campaigns to encourage water conservation.[24] In some areas, the regular water supply was so depleted that water had to be delivered by tanker. Conditions also led to widespread wildfires that damaged natural habitats and led to a decline in crop yields. The drought exposed the vulnerability of the UK's water supply systems to extreme weather events and led to significant changes in water resource planning and management. It was an extremely stressful event for people and trees alike that most people living in the UK more than 50 years ago (including my parents) have vivid memories of. And because of these memories, people of this age have adjusted their behavioural responses when subsequent drought events occur.

Moving beyond record-keeping and in the absence of hippocampi, amygdalae and prefrontal cortexes (core parts of the human central nervous system responsible for memory), did trees also retain a functional memory of the 1976 drought? If they did so, how? How long can such memories persist, given that trees can live for hundreds of years? Do these memories enhance their responses to future drought events? Can they be passed on to progeny via seeds? These are all questions that are being asked by researchers of the UK-based MEMBRA project.[7] I'll discuss this project next, but first, have you ever thought about memory as being anything but a brain-centric phenomenon? And speaking of 'thinking', is cognition a brain-centric phenomenon?

Memory in the traditional sense

What exactly is memory? In a human sense, it's a complex cognitive process that allows organisms to learn from past experiences, adapt to their environment and plan for the future. Memory typically involves three processes: encoding, storage and retrieval.[25]

Encoding is the first stage of memory, in which information from our environment is processed and translated into a format our central nervous systems can understand and use. This information can come through any of our senses – sight, sound, taste, touch and smell – and is transformed into a form of neural code. Once the information has been encoded, it needs to be stored. The brain has an immense storage capacity, with different types of information stored in different parts of it. For example, the hippocampus is involved in storing long-term memories, while the amygdala plays a role in storing emotional memories. Retrieval is the process of accessing and bringing into consciousness the information that has been stored. This can happen consciously, such as when we recall a specific event, or unconsciously, such as when we ride a bike without consciously remembering how to do it.

Memory can be divided into short-term (or working) memory, which holds a small amount of information for a short period, and long-term memory, which can store much larger amounts of information for potentially a lifetime. Long-term memory can be subdivided into declarative (explicit) memory for facts and events and procedural (implicit) memory for skills and habits.[26]

Memory in an embodied sense

A new wave of cognitive neuroscientists is questioning the brain-centric view of memory and cognition, suggesting that this view may be too narrow. An emerging perspective is that memory may not be solely confined to the brain but could involve the whole body in a systemic way.[27] For instance, part of a vertebrate body's immune system appears to have a form of memory: this is why it's called 'adaptive' immunity.[28] Once exposed to a specific pathogen, it 'remembers' the invader and responds more rapidly and effectively when it encounters the same (or similar) pathogen again.

Another relevant concept is the idea of 'embodied cognition', which suggests that our bodies, senses and interactions with our environment play a significant role in shaping our thoughts and memories.[29] For instance, research has shown that physical sensations can influence our cognitive processes and bodily movements can affect memory recall. Moreover, we often discuss the brain and body as separate entities when, in fact, the brain *is* the body (or part of the collective whole). Recently, Anna Ciaunica and other neuroscientists wrote a paper about how cognition should be considered at body or systemic level rather than just brain level.[30] The authors mention that because, ultimately, our thoughts and decisions (and memories) are influenced by a myriad of interactions that occur throughout the body, from tiny cells to whole systems, including our immune and hormonal systems, then cognition should not be confined to one system alone (i.e., just the brain or central nervous system).

This made me think about cognition and memory at the multi-systemic level and the potential role of microbes in thinking. Suppose microbial symbionts (microorganisms that live inside a host) affect animal feeding decisions, sexual and environmental preferences, and desire to exercise, among other cognitive phenomena. Does cognition then not occur at the 'metaorganism' level (the host–microbe conglomerate, which is what each of us is)? If cognition isn't confined to just the brain but is driven by a complex all-body system, which includes a vast array of microscopic organisms, the process of thinking *within* individuals becomes a multi-species event, not just a multi-system one.

I co-wrote a paper about something similar a few years ago, proposing the holobiont blindspot concept.[31] This is the notion that if we fail to understand and acknowledge the influence of our microbial partners living in our bodies on our decisions and behaviour, we may miss a richer explanation for a given behavioural state or a personality trait. This could have important implications for our relationships. For instance, if one person feels less attracted to another or forgets an important date, it could be attributed to the microbes living within.

This concept can also play out in the plant world. We could be missing a richer explanation if we attribute certain plant behaviours or environmental responses to the plant host, but complex plant–microbe interactions actually drive it. If memory in humans and other animals occurs at the system level (or even metaorganism level), could this also be true in trees? In collaboration with its microbiome, could a tree's complex cellular system effectively remember environmental events or exposures (e.g., to pathogens) and mount efficient responses to such events in the future? More research is needed, but memory, much like broader cognition, is more complex than we usually think, and it's unlikely to be confined to just 'animals' or 'brains'. Finally, there's the fascinating realm of epigenetics, which explores how experiences can leave 'marks' or 'tags' on our genes that can be passed down to future generations, effectively a biological memory of past environmental conditions. While these ideas expand the concept of memory far beyond the traditional brain-centric view and into the plant world, they're still being investigated and debated. However, they highlight that memory may be a far more complex, multifaceted and widespread phenomenon than we've typically considered it to be.

Epigenetics and the MEMBRA project

Imagine a theatre production. The script is your DNA, a set of instructions for what happens during the show. But what determines if the actor shouts, whispers or stays silent, which often happens even if this isn't written in the script? That's where epigenetics comes into play.

Epigenetics – a term coined by British biologist C.H. Waddington in the 1940s – refers to changes in an organism's traits that are not driven by alterations in the DNA sequence itself but by modifications on top of the DNA (the 'epi' in epigenetics means 'above' in Greek).[32] Think of these changes as additional directions in our theatrical script that can instruct genes to be switched on or off. One way this happens is through a process known as methylation, where chemical tags called methyl groups latch onto DNA, typically silencing the gene at that spot. Another involves molecules called histones, around which DNA is wrapped. Altering these histones can change how tightly the DNA is wound, thus controlling how accessible certain genes are for reading.

While our DNA sequence is passed down from our parents and remains fixed throughout life, environmental factors such as diet, stress and toxins can influence our epigenome. This helps explain why identical twins, although they share the same DNA, can exhibit different traits and susceptibility to disease – their epigenomes can diverge over time. Perhaps the most thrilling aspect of epigenetics is the idea of transgenerational epigenetic inheritance – the concept that some of your life experiences could influence your offspring's genes.[33] Think about it: the food you eat or the stress you experience might leave an epigenetic 'mark' on the DNA you pass onto your children. Though this area is still being actively researched, and there's much we don't yet understand, epigenetics has already revolutionised our understanding of genetics. It's as if we've been reading the text of our DNA for decades, only to discover that we've been missing a whole layer of footnotes. Indeed, the field shows us that life is not just written in the letters of our DNA but also in the invisible instructions that dictate how the script is performed. It shows us that while our genetic code is important, it doesn't tell the whole story of what makes us who we are. But not just humans have an epigenome – all organisms do. Therefore, trees also have an epigenome and are affected by epigenetics. This is where the MEMBRA project comes into play, as it aims to use epigenetics to explore tree memory.[34] Why is this important? Well, tree memory might be fundamental in allowing trees and forests to adapt to environmental changes. The project was recently awarded £1.8 million to advance the concept of memories of stress in trees. MEMBRA project members have described how previous research has established that plants can write and access a record of prior stress, making the phrase 'memory of trees' "as meaningful to discuss as the selfishness of genes or the (artificial) intelligence of a machine".[35] MEMBRA researchers say that tree 'memory' doesn't rely on genetic alterations in the traditional sense. Instead, it's based on modifications to their DNA's epigenetic signature – adjustments that change the activity of specific genes without altering the actual DNA sequence, as discussed earlier. Picture this as tweaking the volume knob on specific genes. The crucial elements in this are methyl groups, which replace hydrogen atoms in the DNA base, and cytosine, leading to long-lasting memory imprints, including an intriguing phenomenon known as transgenerational resistance. Such revolutionary findings have opened a Pandora's box of possibilities. MEMBRA isn't just talking about enhancing our understanding of tree evolution and resilience but also triggering a profound shift in how we value and respect our leafy companions. This shifts us from a cold, utilitarian view of 'ecosystem services' to a more empathetic understanding that incorporates the idea of tree memory.

MEMBRA is an ambitious initiative that's designed to shed light on how trees 'remember' and how this understanding can influence our ethical and moral perspectives on nature. The MEMBRA team's currently identifying species and populations that exhibit superior resilience and adaptability, using these insights to inform conservation and planting strategies. The project's committed to developing a new language that will weave the concept of tree memory into policymaking. Picture a forest intentionally planted and nurtured, not just for its environmental contributions but its memory-inspired significance. That's the goal of this project – to create the first-ever 'MEMBRA Treescape', a convergence of cutting-edge science, conservation and an ethical understanding of the natural world. In essence, MEMBRA is about understanding, appreciating and using the 'memory' of trees to create a future where humanity and the rest of nature coexist, not just survive. We're talking about a future where every leaf, branch and root tells a story – a memory-filled narrative of resilience and survival. Quite the twist in the tale of trees, wouldn't you agree? Along with cutting-edge science, the MEMBRA project has an artistic side. I spoke with a MEMBRA project member Shelley Castle, who works on the project as an artist. Alongside her, the team enlisted the help of three other creative visionaries to weave their magic into the fabric of the scientific endeavour, partnering with the researchers to produce a blend of art and science bound to ignite curiosity. Think of it as a science experiment but with a twist of creativity. The team's combining the insights from the lab coats with the imagination of artists to communicate the fascinating findings of their research. These artists aren't just painting pictures; they're narrating tales, breathing life into myths and exploring the concept of memory in humans and the awe-inspiring world around us.

The artists hail from different walks of life, each bringing unique perspectives and practices to the table. United by their fascination with interdisciplinary work and forest ecosystems, they form a group known as Walking Forest.[36] They're creative tour guides, who have concocted a 'Creative Engagement Kit' that's packed with fun and engaging activities. This will invite people to immerse themselves in the natural world from the comfort of their homes. It'll guide people to observe closely, unleash their creativity through drawing, delve deep into their thoughts through journalling and weave stories through creative writing. The MEMBRA project is not just exploring the world of trees; they're inviting people to be part of it, to interact, to engage and to appreciate the beauty and complexity of forests through the marriage of science and art. I look forward to seeing how the project progresses.

I recently came across a paper discussing epigenomes in the soil environment. Soil microbiomes, particularly plant growth-promoting microbes, are vital for sustainable agriculture, but understanding how to utilise these beneficial microbes effectively is still challenging. Researchers have inoculated such microbes into the ground to help plants grow. They've found that although these bacteria are often eliminated in soils after their introduction (e.g., killed off by predatory viruses), their presence can still significantly enhance plant growth.[37] The key to this intriguing phenomenon lies in a three-way interaction involving the added bacteria, the native soil microbiome and the plant itself. The study discovered that plant development largely drives the composition of the rhizosphere microbiome, the soil region around the plant root, and that changes to this microbiome aren't essential for promoting plant growth. Instead, introducing the beneficial microbes triggers modifications in the DNA methylation patterns within the plant roots, affecting gene expression and consequently boosting plant growth. These epigenetic DNA changes continue to influence growth even after the bacteria are removed from the soil. This research unveils a novel mechanism of plant growth promotion, wherein the inoculated beneficial microbes alter root DNA methylation, providing valuable insights for future strategies in plant microbiome engineering. Therefore, it seems that below-ground epigenetic changes in trees are important to their survival. I now wonder if the soil microbiome plays a role in a tree's memory. Hidden from view but never fading from significance, epigenetics could enrichen our understanding of memory and bring a new appreciation for other-than-human species.

Memory and cognition are far more complex than we're taught at school. They involve multiple systems (and perhaps multiple species, visible and invisible, working as one), and epigenetics adds another layer of complexity. Yet we shouldn't shy away from complexity. It's an inherent part of life, and its richness often leads to growth of mind. When we avoid complexity, we might miss opportunities for learning, development and the chance to develop empathy, and even to discover novel solutions. Embracing complexity allows us to see the world in a more nuanced way. It pushes us to adapt, learn new skills and develop a more comprehensive understanding of the world around us. In essence, complexity breeds resilience, sparks creativity and encourages a deeper appreciation of the interconnectedness of things. By tackling complexity head on, we equip ourselves to navigate challenges better and foster a more inclusive future. Trees can teach us these things.

And thus spoke the plant...

I discovered Monica Gagliano's work around six years ago.[38] When I started my PhD, I was really interested in pushing the boundaries of current knowledge, and Gagliano's work struck a chord. She's an Italian-Australian ecologist and environmental scientist known for her work in the field of plant cognition and communication. She's become famous for her studies on the ability of plants to learn behaviours and communicate in ways previously thought impossible for organisms lacking a central nervous system. Early in her career, Gagliano's research focused on animal life, specifically the impacts of anthropogenic noise pollution on fish behaviour and physiology. Over time, however, her interests shifted towards a broader understanding of life, which ultimately led her to the field of plant biology. Specifically, she became intrigued by the idea that plants might be capable of learning and remembering information, similar to animals. This represented a significant departure from traditional scientific beliefs about plants, often considered passive organisms.

Gagliano started making headlines with her research on plant behaviour and cognition in around 2014. Her experiments on *Mimosa pudica*, often known as the 'sensitive plant', were particularly significant. She found that it could learn and remember information, challenging the conventional scientific understanding of plant behaviour.[39]

The *Mimosa pudica* plant is known for its dramatic physical response to touch; it folds its leaves inwards when disturbed – whether this is a light touch, a shake or exposure to heat. This phenomenon is known as thigmonasty or seismonasty. I vividly remember caressing the leaves of the plant at the Eden Project, in Cornwall over a decade ago (and more recently in the jungle of Kadavu, Fiji). I can still picture the plant's reaction and my reaction of amazement to it. The folding motion occurs in a matter of seconds, and is considered to be a defence mechanism to help deter herbivores: the movement can make the plant look less appealing or startle a potential predator. The plant can also use this mechanism to protect itself against environmental stressors such as heavy rainfall. The folding action is due to a rapid change in turgor pressure within the plant's cells. The stimulus triggers an electrical signal, which results in the loss of potassium ions in the cells at the base of the leaflets, causing water to follow by osmosis and leading to a rapid change in the cell's turgidity: this results in the closing motion.

Mimosa pudica *leaflets, which close up like a book when touched.*

The sensitivity of the *Mimosa pudica* plant isn't limited to its leaves. Its stems can curl up when touched, and the plant will fold its leaves at night as it enters a 'sleep' state.

In one of her experiments, Gagliano tested whether *Mimosa pudica* could learn to differentiate between harmful and non-harmful stimuli. The experiment involved dropping the potted plants about 15 cm. This height was chosen because it was high enough to stimulate the plant's defensive response but not high enough to cause actual harm. Over a series of 60 drops, Gagliano found that the plants initially closed their leaves, as expected. However, over time, they began to keep their leaves open despite the repeated drops, indicating that they had learned the stimulus was not harmful. Even more remarkably, the plants retained this learned behaviour several weeks later, suggesting a form of memory. This was a ground-breaking discovery, as learning and memory were traditionally thought of as capabilities unique to organisms with a complex nervous system. This experiment challenged long-standing beliefs about plant behaviour and cognition, suggesting that plants could learn from their experiences and adapt their behaviour accordingly. Gagliano's work on this has been both celebrated and controversial, with some scientists questioning her methods and conclusions. However, it's undeniably contributed to a broader conversation about our understanding of plant life, leading to more research on plant cognition and communication. Gagliano's recent research has included areas such as plant bioacoustics and the impact of sound on plant growth. One of her notable studies in this field involved the garden pea (*Pisum sativum*).[40] In a series of experiments, she showed that these plants can sense the vibration of water moving through a pipe even without physical contact. The plants appeared to use this sound to orient their growth towards the source. In the experiment, Gagliano set up 'mazes' for the pea seedlings with a single water source in one arm of the maze. Some mazes used a pipe to deliver water, creating a vibration that the plants could potentially 'hear'. The pipe was present in other mazes, but no water flowed through it. She found that the plants were more likely to grow towards the arm of the maze with the vibrating pipe, suggesting they could sense the sound of the flowing water. This work suggests that plants can perceive their environment and react in more sophisticated ways than previously thought, including potentially 'hearing' or responding to sound.

A recent study in the journal *Cell* strongly aligns with Gagliano's theories. Titled 'Sounds emitted by plants under stress are airborne and informative',[41] it demonstrates that plants emit ultrasonic airborne sounds, effectively 'verbalising' their stress. This revelation opens exciting avenues in plant–environment interactions research, as these acoustic emissions can be harnessed to train machine learning models to detect and identify plant stress, such as dehydration and injury. This development paves the way for numerous future applications and holds great potential for advancing our understanding of the intricate relationship between plants and their surroundings. So we now need to view plants as 'soniferous' (sound-producing) organisms.

Throughout her career, Gagliano has faced significant pushback from some in the scientific community owing to the unconventional nature of her work, which challenges long-held beliefs about the capabilities of plants. Some scientists have questioned the methods she's used in her experiments. They argue that her studies lack sufficient controls or that her interpretation of the results is overly anthropomorphic (attributing human characteristics to non-human entities). For instance, sceptics might argue that other, simpler biological mechanisms could explain the behaviours she interprets as learning or memory. They contend that these characteristics are inappropriate to apply to organisms that lack a nervous system or brain, and that such attributions risk misleading the public's understanding of plant biology. Any time established scientific beliefs are challenged, it can lead to controversy and criticism, as it takes a while for new ideas to be thoroughly vetted and accepted by the broader scientific community. Despite the criticisms, it's important to note that science is a field that thrives on scepticism and rigorous testing of new ideas. Gagliano's work, while controversial, has sparked important discussions about plant cognition and communication and has opened new research avenues.

In their underappreciated majesty, trees (and other plants) can sense their environments, the rumbles of water and the vibrations of other creatures. They probably even remember, capturing and storing the echoes of bygone events and responding via epigenetic mechanisms and ways we don't fully understand. Take a moment to think about how the wooden bed you lie on, the chopping board you slice your vegetables on, the coal you burn and the paper you write on might once have had senses and memories.

Industrial Chemicals – I Once Was a Tree

Many industrial chemicals, crucial to various manufacturing processes, are derived from wood and other plant materials. Three notable examples are methanol, acetic acid and formaldehyde. Methanol, often called wood alcohol, can be produced through destructive distillation of wood. When wood is heated without oxygen, it decomposes into its components, including methanol. This versatile chemical has numerous applications, from being a solvent in paints and varnishes to a building block for producing plastics and synthetic fibres.

Acetic acid, commonly found in vinegar, can be obtained through the fermentation of wood products, such as wood chips or sawdust. Acetic acid finds its way into various industries, including producing textiles, solvents and food preservation.

Formaldehyde, another important industrial chemical, can be derived from wood by a process called wood pulping. Wood pulp is treated with various chemicals, leading to the release of formaldehyde. This compound is widely used to manufacture resins, plastics and adhesives.

CHAPTER 15

Dark Emu: An Indigenous Perspective

"If we are to attempt to understand Indigenous philosophy, it has to begin with the profound obligation to land."
—Bruce Pascoe

Within the diverse troves of human knowledge lies a deep wisdom forged over millennia. A set of knowledge systems that are often dismissed. A collection of worldviews that are often denigrated with words such as 'pseudoscience'. However, this is a misunderstanding of the nature and value of Indigenous knowledge. This is a complex set of diverse systems of understanding that encompasses scientific knowledge and more subjective spiritual and cultural dimensions, particularly regarding how our beingness is shaped by and with the rest of the natural world. These systems and perspectives are often overlooked by modern science. One of the problems with this kind of knowledge is brought about by a lack of consensus on the notion of different types of knowledge acquisition.

Different disciplines, cultures and philosophical traditions have recognised various forms of knowledge acquisition, from perception (knowledge gained through our senses) and reasoning (using logic and critical thinking to generate knowledge) to intuition (knowledge gained through unconscious reasoning). There's also an added complexity when considering broader philosophical camps, with different ontologies (the study of being) and epistemologies (the study of knowing).

Take positivism, for example. This philosophical approach falls under the umbrella of epistemology. Positivism holds that the only authentic knowledge is based on experience and positive verification.[1]

This approach emphasises empirical evidence (data collected through senses and direct observation) and the use of scientific methods. It asserts that all genuine knowledge can be reduced to logical and mathematical treatments. Therefore, it rejects intuition, metaphysical explanations and other forms of knowledge that can't be empirically verified. However, positivism is not without criticism. Some people critique it as overly simplistic, arguing that it ignores the complexity of human experience and the subjective aspects of human knowledge.

By dismissing Indigenous knowledge as invalid or pseudoscientific, we disregard the vast wealth of knowledge accumulated over generations. But we also perpetuate a harmful narrative that only one way of acquiring knowledge is valid. This can erode the cultural identity and autonomy of Indigenous Peoples and undermine the potential of solving complex problems (such as deforestation) through collaborative approaches.

A note on management

The modern world is obsessed with management, managing and managers. I haven't forged a solid opinion on this yet, but I'd like to take this as an opportunity to think it through. An obsession with management seems to reveal our preoccupation with order, efficiency and productivity, often at the expense of creativity, spontaneity and collectivism.

I think an emphasis on management reflects many people's belief that everything can and should be quantified, measured, and controlled. It assumes that all aspects of the world are predictable, knowable and controllable, and that everything can be reduced to a set of processes and procedures. However, while management and quantification can bring considerable benefits in many situations, they overlook the complexity and unpredictability of human nature and the natural world. Moreover, our obsession with management often creates a hierarchy of power and control that reinforces existing social, economic and political structures. But rather than being inherently bad (I need some management in my life), I think the risk of dominion is the real problem with management, particularly when it comes to more-than-human nature.

Dominant management paradigms often prioritise economic growth and short-term profits over long-term sustainability and the wellbeing of communities. This goes hand-in-hand with the exploitation of natural resources, loss of biodiversity, and degradation of ecosystems, including forests. Perhaps management is not the issue – it's the worldview that cradles it. Historian Bill Gammage writes: "Management assumes hierarchy – people manage, the rest are managed. 'Care' is closer to the Indigenous perspective ... But this, too, assumes hierarchy. 'Collaboration' is the closest. All creation shared responsibility for maintaining the universe. This was the point of *Ceremony*. Ceremony asserts the interdependence of life and things."[2] While it could be seen as a semantic and somewhat contradictory issue, I respect the view that our relationship with nature should be founded on care and collaboration rather than hierarchy.

Bruce Pascoe

Bruce Pascoe is an Aboriginal historian and prolific author, of Bunurong, Yuin and Tasmanian heritage. He wrote the hugely successful *Dark Emu* and other books on Aboriginal history, rebutting colonial myths, particularly about land management (I use the term 'management' henceforth, but remember, to many Indigenous Peoples, it's more akin to 'collaboration').[3] *Dark Emu* presents evidence to show that Aboriginal peoples in Australia had complex agricultural practices, including sophisticated systems of land and resource management, long before the arrival of European colonisers. Pascoe argues that Aboriginal peoples held a deep understanding of the land and how to cultivate it sustainably; they used various techniques, such as irrigation, cool burns, crop cultivation and seed harvesting, to create a diverse and productive landscape.

One of the most surprising things about the book is that people are surprised by the notion of Aboriginal peoples having used sophisticated land-management practices for millennia. This feeling of surprise is telling. It tells us that the structures and contours of today's mindset were firmly created by centuries of colonial narratives portraying Indigenous Peoples as inferior and lacking in knowledge and skills. These narratives have been used to justify the dispossession and marginalisation of Indigenous Peoples from their lands. Indeed, as Pascoe writes, "the underestimation of Indigenous achievement was a deliberate tactic of British colonialism".

I've spoken with many Australians who tell me true Aboriginal history was utterly absent from their childhood school curricula. Furthermore, most of them are under the age of 40, indicating that this isn't an issue solely confined to past generations. However, *Dark Emu* is now an optional resource in some Australian schools. The omission of true Aboriginal history from curricula has perpetuated the dominant narrative of Indigenous Peoples as primitive and uncivilised. It also reinforces the idea that Western knowledge-acquisition and cultural systems are superior. This erasure of Indigenous knowledge and history has significantly affected Indigenous communities. It's contributed to the marginalisation of Indigenous Peoples from decision-making processes about their lands and perpetuated a cycle of dispossession and displacement that has lasted for centuries.

This is why Pascoe's work is so profoundly important. It's been instrumental in kindling conversations about the history and culture of Indigenous Peoples in Australia, as well as about sustainability and the environment. His work shows that Indigenous Peoples often have a deep understanding of the environment and the importance of sustainable land practices (including conserving and restoring forests) and that this knowledge, when used ethically, can create a more sustainable future for all.

I actually moved to Australia at around the time I started writing this chapter. I decided to try my luck at speaking with Bruce. I'd considered getting in touch with him before – when I lived in the UK – but I held off for several reasons. For one, I felt the contextual relevance would be stronger after I moved to Australia. I first spoke to Bruce about his ongoing experiments growing native foods, such as grasses and tubers, on his farm in East Gippsland (New South Wales). Aboriginal peoples have been sustainably producing diverse foods from their native ecosystems for thousands of years, including possibly the world's oldest bread. Yet many of these foods and the knowledge of their cultivation are at risk of being erased owing to social structures steeped in colonial history. Bruce is trying to reintroduce and maintain these native foods. I asked Bruce how Indigenous Peoples in Australia traditionally managed and restored forests and how we can learn from these practices today. He said, "It's hard to provide a simple answer to that – knowledge of how these practices were conducted is fragmented, but we can talk about how it was and continues to be maintained. Our people often established a mixed forest and cropping system and maintained it with fire. Unwanted trees in areas where crops grew were removed." Bruce then talked about running two trial plots on his farm in Gippsland. In one plot, the forest-crop system results from continuously logging around a third of all trees. This stimulates a lot of regrowth while providing space for food crops. In the other experimental plot, he keeps tree removal to a minimum, establishing what he and his collaborators believe it was like before human intervention. Bruce is in contact with other Aboriginal organisations undertaking similar trials to study how forests were restored and maintained. They hope to share their results in 20 years. I asked Bruce if he could speak of some of the roles that forests play in Indigenous culture and spirituality, and how this shapes Indigenous approaches to forest management. He said, "Our peoples generally have a deep spiritual and philosophical connection to the forest. In essence, we see Earth as our mother and treat her accordingly. This is not just an affirmation. Treating Country in such a way is deeply ingrained in our values and belief systems." He then spoke of the Creator Spirits. It's believed that Aboriginal Creator Spirits reside in the Dreaming or the 'Spirit World' that accompanies our physical one. The Dreaming is the basis for all the values and beliefs Aboriginal peoples choose to live by. It's perhaps worth defining 'Country' here. From an Indigenous perspective, the term refers to more than just a geographical or political territory. It encompasses a holistic and interconnected understanding of the land, its physical features, ecosystems, spiritual significance, cultural heritage and ancestral connections. Country is deeply rooted in Indigenous cultures and is intricately tied to traditional knowledge, customs and spiritual beliefs. It forms the foundation of Indigenous identity and provides a framework for land stewardship and sustainable relationships with the natural world.

Bruce continued, "Stories in the south coast talk of the Creator Spirits, which gave people two great gifts – one was the rock, and one was the tree – and everything needed could be found in these two things." This notion holds a profound truth. After all, rocks and trees are foundational to soil formation and food systems. Much of what we need and feel we need is deeply tied to trees. "What we can learn from this today is simply that we must look after the land, air, and seas! Whatever we do must have sustainability at its heart," Bruce remarked. I pointed out that in his writing he had highlighted the importance of understanding Indigenous knowledge and practices in the context of land management. But I wanted to know how this knowledge could be integrated into contemporary approaches to forest restoration. Bruce told me that one profoundly important Aboriginal land management tool was and still is cool burning. You may recall that I mentioned this in Chapter 2.

Bruce mentioned that he wrote a book with Bill Gammage called *Country*, which details cool burning and Aboriginal land management.[4] This provides a roadmap for moving away from the destructive land management practices that still dominate Australia and other areas of the world. As Pascoe said, it's as if the colonists "went straight to replicating the systems of the lands they had left, as if they had moved next door, not to another hemisphere". These destructive land management practices leave the Australian forests vulnerable to uncontrollable wildfires. The book also offers insights into how the animate and inanimate are inseparable in the Aboriginal concept of 'Country'. As Margo Neal says in the foreword, "In this worldview, everything is *living* – people, animals, plants, rocks, earth, water, stars, air and all else." Reciprocity with the land is at the heart of this thinking – if you care for the land, the land will care for you.

One of the key messages I took from this book was, as Bruce emphasised during our chat, that fire was a primary tool for farming and sustainable forest management. Aboriginal peoples grow food using fire as one of the productive management methods – they burn regularly. Bruce is highly involved in cool burns today. He told me he'd been burning regularly as there were currently no local fire restrictions. He then mentioned that he and his collaborators run workshops on fire management using cool burning techniques. Aboriginal peoples are rightly prioritised in the workshops, but they're still open to non-Indigenous folk.

I asked Bruce how the colonisation of Australia has impacted the health and resilience of forests and what steps could be taken to restore them. He said, "It's not just the colonisation of people. The unsustainable agriculture that came with the Europeans has warped the forest's make-up. For instance, huge numbers of sheep ate the recruitment trees of the wattles and eucalypts, so the forest types have dramatically changed. Wildfires (often caused by poor land management) also meant the understorey was emergent and much thicker than before. Without the restraint of cool burning, it's not a good situation – as highlighted by the devastating 2019 fires." He told me that modern forestry methods also create pencil-thin trees, exacerbating the fires.

And the way to heal the land? Well, promoting Indigenous leadership and sovereignty and funding Indigenous sustainable practices is one important road to healing for the people and the land. Stopping unsustainable forestry and agriculture is another.

I wanted to know what advice Bruce would give to non-Indigenous Peoples and organisations interested in supporting Indigenous-led initiatives for forest restoration and management. I, for one, am keen to contribute, but knowing where to begin is challenging. Bruce spoke just three words, "Learn Australian history", then paused. "Learn as much as possible about the country's past and present state. This is a good starting point," he continued. "The early colonists wrote how not only could they see through the horizon because the trees were so large, but they could also ride a horse in a straight line through woodlands, indicating the Aboriginals' sophisticated land management. Learn the *real history*. We also need to engage in better conversations. Indigenous and non-Indigenous Peoples need to come together without knee-jerking and jousting. We need to work together now to co-create the best practices for Mother Earth."

I asked if he thought the modern social media culture was pouring fuel on the fire – stopping us from working together. He said, "Yes, although it provides the benefit of better connecting our minds and voices, it does seem that through social media, the uninformed voices hold more sway than those with deep knowledge and wisdom." I moved our conversation in a slightly different direction. "What are your views on balancing economic development with the conservation and restoration of forests, particularly in the context of Indigenous land rights and sovereignty?" Without a second of hesitation, he replied, "If we don't look after the Earth, there will be no economy!" A short but perfect answer. "In *Dark Emu*, you challenged the prevailing narrative around Indigenous Peoples as 'hunters and gatherers', instead highlighting how they were actively cultivating and managing the land before 1788. How can this understanding inform our approaches to forest restoration and management?" Bruce replied, "Australian people must understand that the history taught to them was incorrect, which inhibits sustainable practices today. Suppose non-Indigenous people understand that Aboriginals were and are still intellectual land managers and actively work to promote Indigenous leadership. In that case, we can all learn how to restore and manage our forests and live in harmony with the land and each other."

I've had the privilege of working with several Indigenous scholars in my academic career. One of the most important things I've learnt is that Indigenous language and knowledge are intimately connected with biodiversity and the land. For instance, in 2021 I co-wrote a paper called "Traditional ecological knowledge: a call to listen deeply, engage with, and respect Indigenous voices".[5] For this paper, along with my non-Indigenous coauthors, I worked with talented Indigenous scholars Kim O'Donnell and Nicole Redvers, who enlightened me about the interconnections between Indigenous languages and local biodiversity. Traditional ecological knowledge is an expression of the deep spiritual and historical relationship that Indigenous Peoples have with their land. This knowledge is typically shared verbally within communities and passed down through generations via cultural traditions such as storytelling, song and ceremony. Indigenous languages are closely tied to the ecosystems and landscapes in which they exist. Thus, when there's a loss of biodiversity or restrictions on land use, the practice and transmission of this traditional knowledge and language are greatly hindered or lost altogether. Ultimately, these losses pose a significant threat to Indigenous Peoples, as the erosion of their culture and language is intimately tied to the erosion of the land itself.

In our paper, we argued that forest restoration projects that do not consider the needs of Indigenous communities are more damaging than restorative. This is because the land is often viewed as a 'relative' by Indigenous Peoples. When they cannot access or interact with the land freely, they lose their reciprocal teachings and relationships with it. This also applies to Indigenous languages. As biodiversity declines and access to land is reduced, traditional botanical knowledge is lost and languages disappear. This puts traditional knowledge and languages at risk. We called for its revitalisation and the immediate return of land tenure rights. The forest (and other systems) restoration community can play a role in this process by becoming humble allies and engaging in community actions prioritising cultural safety and responsiveness.

To be truly restorative, one must have the capacity to renew continually, and so-called traditional ecological knowledge has the language and tools to maintain this potential over time. Creation stories of Indigenous Peoples embody direct relationships between all elements of the universe. They are rooted in a sense of spiritual embodiment and natural harmony – an ecocentric rather than ego-centric view. This contrasts with colonial ideals and mindsets that have dominated. For example, the Maralinga Tjarutja People have sustained themselves on the Ooldea soak (a spring in South Australia) for 60,000 years. In contrast, it took only 60 years from colonisation for colonial practices to ruin this precious water source in Australia's desert outback.

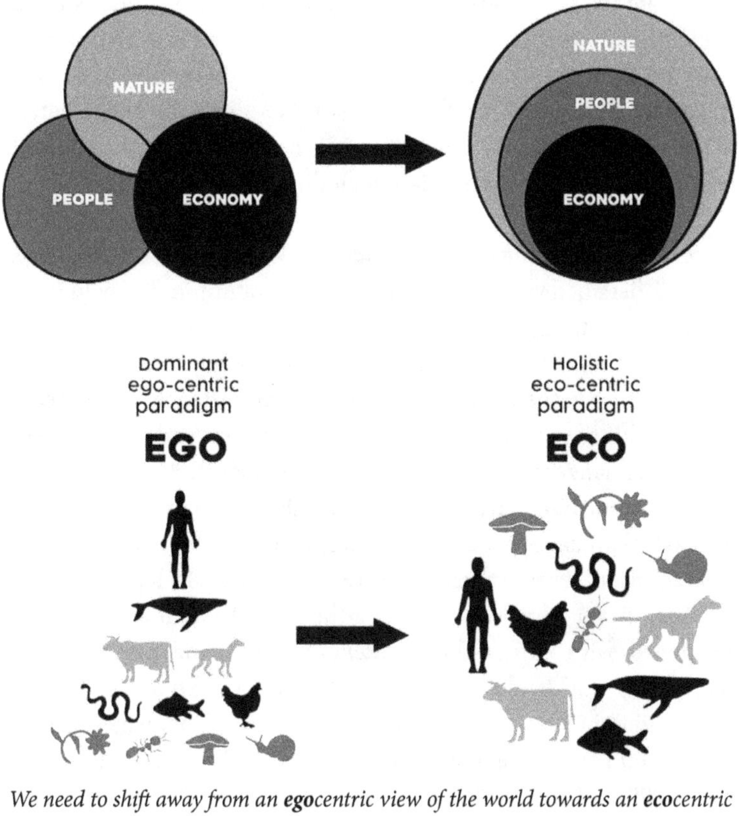

*We need to shift away from an **ego**centric view of the world towards an **eco**centric view if we're to stop spiralling climate change and biodiversity loss. (Artwork designed by Barkindji-Malyangapa artist Jasmin Craciun.)*

I wanted to know how people can support the revitalisation and preservation of Indigenous languages and knowledge systems, which are often closely tied to forest and land management practices. Bruce said, "This is also a complex issue, but money needs to be spent on Aboriginal language research. I'm on the board of *First Languages Australia*, which encourages communication between communities, the government and key partners whose work can impact Aboriginal and Torres Strait Islander languages. And to be honest, they struggle for oxygen in this debate." By this figurative expression, I think Bruce means that Indigenous viewpoints are not receiving enough attention or consideration in the wider discussion or debate. Their arguments are being suffocated by opposing viewpoints, making it difficult for them to be heard or properly considered. Bruce went on to say that we need more modesty and proper engagement in conversations about language conservation and extinction. "One problem is that universities are hunting research dollars because of how they have been privatised. But one of the focus areas needs to be the conservation of Aboriginal languages, which, because they are interdependent, will benefit biodiversity conservation, including our precious forests."

I decided to investigate the extinction of Indigenous languages in more detail, reviewing the scientific literature and trawling the internet for data on current languages and historical language losses. It turns out that languages are categorised much like biodiversity when it comes to their conservation status. For instance, some species are flourishing, some are threatened and some are endangered (even critically endangered). The same goes for languages.

The astonishing part, to me, is that the current extinction rate is around 90 languages per decade. Ninety languages that take countless generations to develop and evolve are vanishing every ten years, gone and often never to return. At least a third of all extant languages are threatened with extinction. Incidentally, the same proportion of tree species worldwide are also threatened with extinction. Indeed, it's thought that by 2080, the extinction rate of languages will be around 160 per decade, rising to 260 by 2150.[6] This means we might lose 50–90% of the world's existing languages in the next century.[7] Just as we need to protect trees and other biodiversity, we must do something about this shocking trend. Ancient diversity within human populations is diminishing, and with this loss comes a loss of nature and vice versa.

Recently, I decided to collaborate with Ochre Dawn and the Indigenous Literacy Foundation, Indigenous neuroscientist Araceli Camargo at the Centric Lab and psychologist Miles Richardson to raise greater awareness about this matter. We created the Language Stripes to show how Indigenous language extinction rates are increasing over time. Many organisations are raising awareness of this issue and working on conserving languages, such as the Endangered Languages Project and the Language Conservancy, to name a couple.[8]

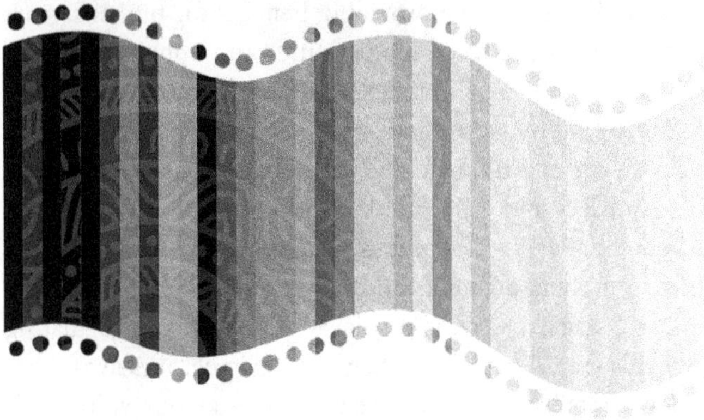

The Language Extinction Stripes, showing language declines from 1700 (far left) to 2150 (far right).

Thus, conserving Indigenous languages is conserving biodiversity (including trees) and vice versa.

Reciprocal restoration

Reciprocal restoration is a concept that goes beyond traditional restoration practices, which often focus solely on repairing damage to natural systems caused by human activities. Instead, it aims to establish a two-way relationship between humans and the environment, with both benefiting from the restoration process.[9] This approach recognises that humans are a part of nature and that our wellbeing is intimately connected to the health of the ecosystems that sustain us. At the time of writing, I'm working in the Frontiers of Restoration Ecology (FORE) Lab in Adelaide, South Australia.[10] The lab's led by Martin Breed, and the team has an ethos steeped in the reciprocal restoration approach. One of our research strands focuses on understanding how to restore ecosystems for both the ecosystem's intrinsic value and the health benefits it brings to humans. An example of reciprocal restoration we're working on is the restoration and improvement of urban green spaces. Many cities worldwide are grappling with issues such as air pollution, urban heat islands and limited access to green spaces and biodiversity that bring myriad health benefits. This lack of access to quality environments can negatively impact human health and wellbeing.

By restoring and expanding urban green spaces, we can protect and enhance biodiversity while providing benefits for humans, such as improved air quality, reduced temperatures and opportunities for outdoor recreation and social interaction. There's also an important biological connection between biodiversity and the human immune system. We're living in a time where two global megatrends – a rise in immune disorders and a decline in biodiversity – are intimately connected. The 'biodiversity hypothesis' suggests that exposure to a wide range of microorganisms from our surroundings, particularly during the early stages of life, is crucial for training and regulating our immune systems. Therefore, when we distance ourselves from biodiverse environments, we inadvertently increase our vulnerability to inflammatory diseases. As such, we're trying to improve urban green spaces and their microbial residents to enhance biodiversity and human health. We know that improving the tree coverage in our cities can enhance the diversity of microbes in the air and potentially reduce the relative abundance of pathogens. This is another reason to conserve trees. But reciprocal restoration goes further than just creating green spaces. It also means engaging and empowering local communities in the restoration process. By involving community members in the planning, design and implementation of restoration projects, we can build social cohesion, create job opportunities and foster a sense of ownership and stewardship of the restored areas. This certainly aligns strongly with my conversation with Bruce Pascoe. Restoring and enhancing natural systems can support human wellbeing and social equity and create a more sustainable and resilient future for ourselves and the planet.

Plastics – I Once Was a Tree

Polyethylene is a versatile and widely used polymer (a large molecule of long chains of smaller molecules) belonging to the polyolefins family. It's a type of plastic that can be melted and remoulded multiple times without undergoing significant chemical changes. Polyethylene is known for its combination of properties, which include toughness, flexibility, chemical resistance and low cost. Imagine tiny building blocks called ethylene molecules, consisting of two carbon atoms and four hydrogen atoms (C_2H_4). These humble units come together in a process called polymerisation, in which thousands of ethylene monomers chemically link up to form long chains.

Polyethylene plays a starring role in various industries, from packaging and construction to automotive and consumer goods. Its lightweight nature, durability and ease of processing have made it a favourite among manufacturers looking for a cost-effective material. While most polyethylene today comes from fossil fuels such as crude oil, scientists are actively exploring ways to produce it from renewable plant-based sources, opening the door to a greener future. The extraction of fossil fuels for polyethylene production contributes to environmental pollution and poses a threat to natural habitats. It's a sobering realisation that the same materials once derived from trees are now used to create polyethylene, which can inadvertently harm living trees. From one perspective, we're using dead trees to kill living trees.

CHAPTER 16

Restoring Forests in a Changing Climate

"We must accept finite disappointment but never lose infinite hope."
—Martin Luther King Jr

The Earth, our shared cradle and home, bellows out a song of warning. The strident melody weaves through dust clouds and rustling leaves, echoes in the retreating glaciers and rides the ever-strengthening storms. Many of us feel this rhythm of disruption. We stand on the threshold of a new era, our actions reverberating through the atmosphere, stirring the delicacy of life. Climate change has cast its long shadow over our world, reshaping the lands, waters and skies. It's a metaphorical chapter that we, as humans, are coauthoring.

The tale is one of fire and ice, of species lost and habitats altered. Will we now lead with respect and reciprocity or stumble in ignorance? The next steps are ours to choreograph.

As I write this, I'm sitting outside watching the morning sun – the great giver and taker of life – peek over the horizon, casting a mosaic of light and shadows on our troubled tableau. A question that springs to mind, one that so often does, is how will our epochal chapter unfold in these unparalleled times? I look at the silhouettes of trees around me. Slowly, as the sun's rays hit their needles and leaves, the trees glisten, triggering a cascade of neuronal activity and emotions. I'm reminded of the beauty, power and interconnectedness of the elements within our biosphere and the immense astral beacon that drives it all the way out there. Yet I also find myself pondering the fate of these glistening trees in our changing climate. Whether it's brought about by misinformation, psychological distance, a perceived lack of immediate effects or other reasons, climate change is still an underappreciated phenomenon that dramatically affects our precious forests (and other ecosystems).

The composition of a forest is not fixed – it always has been and always will be dynamic. But a wave of discordancy is altering it at an unprecedented pace and magnitude. Climate change causes unpredictable and extreme events, such as forest fires and floods, which can have an unfathomable negative impact on forests and their restoration. However, climate change can also alter the general ecological condition of a forest, leading to droughts, diseases and changes in plant communities.[1] Warmer temperatures can increase the rate of tree growth in some regions, but in others it can exacerbate water stress and lead to reduced growth. Similarly, changes in rainfall patterns can affect water availability, with drought conditions making trees more susceptible to disease and death.[2] Warmer conditions can enable disease carriers to expand their range and increase their survival during winter. This can lead to larger and more frequent outbreaks, causing significant damage to trees. For instance, bark beetle outbreaks in North America have been linked to warmer temperatures, allowing the insects to reproduce faster and survive in regions previously limited by cold.[3]

Shifts in species distributions

In the USA, I once had an encounter with a bird that almost led to my arrest. It's called the northern cardinal, known colloquially as the redbird. The cardinal is a striking and iconic bird species. The male is known for its vibrant plumage, with a brilliant red body and a prominent crest on its head. Its wings and tail display accents of black. In contrast, the female is predominantly greyish brown with reddish tinges on her wings, crest and tail. Both male and female cardinals possess a distinctive black mask around their eyes. They have thick, cone-shaped bills, perfect for cracking open the seeds and fruits that form most of their diet. Their strong beaks allow them to access a wide range of food sources, including insects and berries. They're non-migratory birds, often seen year round in their chosen habitats, which include woodlands, gardens and thickets. They evolved to prefer woodland edge habitats.

The beautiful northern cardinal.

I spotted the bird while attending a planetary health conference in the USA several years ago and began researching its ecology. The event was held in Detroit, Michigan. Before the conference, my wife and I ventured out into a patch of scrubby habitat with a pair of binoculars after we saw the bird hopping around in the scrub from our hotel window. Within three minutes, we heard a van pull up, seemingly out of nowhere, and several police officers jumped out, demanding to see our ID and questioning our presence. Admittedly, we looked a little shady, but this felt like an overreaction of immense proportions. We turned around with binoculars in our hands, which seemed to unnerve the officers. I said, "We're just looking at a bird," feeling like a total nerd. One officer said, "Oh yeah," as if he didn't believe me, "What kind of bird?" "Northern cardinal," I said quickly. After several minutes of trying to convince them we were just a couple of

Brits appreciating the beauty of the local passerines, they let us get on with our day. Needless to say, it was an amusing story to tell at the conference dinner.

So, how does the northern cardinal relate to climate change? Well, as temperatures have warmed, these birds have been observed expanding their range northward into regions that were previously too cold for them to inhabit. This range expansion is believed to be facilitated by milder winter conditions and the availability of suitable habitat and food resources – they love seeds, leaf buds, flowers and many berries, wild fruits and insects. Specifically, the cardinal is spreading northwards into Canada, whereas a century ago its latitudinal limit was just north of New York.[4] Similar shifts in species distribution are occurring across the globe, with some expanding their range as their habitat increases, while others become threatened as their habitat dwindles. As temperatures rise, the optimal habitat for many tree species is shifting poleward and to higher latitudes. This can lead to changes in forest composition, with some species becoming less common or disappearing entirely from some regions and other species moving in to replace them. I recently came across a study entitled 'Climate change, tree demography, and thermophilization in western US forests', led by Kyle C. Rosenblad.[5] The study uncovered warning signs indicating that forests in the Western USA are having difficulties adapting to the swiftly altering climate. Rosenblad explained, "If you're concerned about forest health, then one thing you want to observe is whether the rate at which forest composition changes is roughly equivalent to the rate at which the climate changes. We found that climate change is outpacing the ability of forests and tree biomass to keep up, which is a potential concern for forest managers."[6]

Indeed, their study revealed that over the past decade, forests have increasingly been taken over by trees more suited to endure hotter and drier climate conditions, a phenomenon known as thermophilisation. The term thermophilisation comes from two roots: 'thermo' and –'philisation'. 'Thermo' is derived from the Greek 'thermē', meaning 'heat'. It's a prefix often used in scientific terms to denote heat or temperature. Philisation comes from the Greek 'philia', which means 'love' or 'affinity for'. When used in the context of a biological or ecological term, philisation often suggests a shift or adaptation towards the condition specified by the first part of the word. So thermophilisation essentially means a shift or adaptation towards heat, indicating that organisms (in this context trees) adapt to or become more suited to hotter conditions. While the exact elements influencing thermophilisation can vary among different forests, the shift in forest composition is generally not matching the pace of climate change. Water availability is often constrained in many ecosystems and elevated temperatures can accelerate water loss in trees. It's also important to note that extreme elevations in temperatures can kill organisms. As American author Jeff Goodell recently said, "Because all living things share one simple fate: if the temperature they're use to – what scientists sometimes call their Goldilocks Zone – rises too far, too fast, they die."[7]

When a tree's water supply reaches alarmingly low levels, which signifies severe drought stress, its circulatory system (analogous to human veins and arteries or a city's water supply infrastructure) experiences intense strain, leading to the formation of air bubbles that can damage portions of the system, much as the strain on a city's water infrastructure can cause pipe bursts or pump failures. Some trees have evolved to build specialised vascular structures that inhibit bubble formation and propagation as a defence mechanism. Additionally, the leaves of trees can wilt owing to drought stress. To combat this, some trees close the pores on their leaves to reduce evaporation or accumulate solutes in their leaf cells, making it physically harder for water to escape.[8] This is a bit like a city enacting water conservation measures, such as limiting sprinklers, encouraging shorter showers or storing water in containers. Every ten years, Forest Service scientists in the USA measure sections of trees and gather data from across the country, providing insights into the state and wellbeing of the country's billions of trees. This is what is known as a tree census. They can monitor changes over time by assessing and recording tree numbers, types and health. The western half of continental USA is home to nearly 100 different tree species, including the quaking aspen, ponderosa pine and Douglas fir, and nationally there are over 800 native trees.[9] Each species has its own geographical distribution, and a trend highlighted in Rosenblad's study reveals that these species often find it challenging to adapt to climate change in the warmest, driest parts of their ranges. The quaking aspen serves as a prime example of a species struggling significantly with climate change. However, diseases (which are often exacerbated by climate change) and other factors are also contributors.

The research team examined close to 45,000 forest subplots recorded in the most recent tree census. Their analysis discovered that from 2011 to 2020 the temperature in Western forests had risen by an average of 0.57 degrees Fahrenheit. However, it's worth noting that the forest thermophilisation rates lagged, being ten times slower than the mean temperature increase. Heat-induced drought stress in trees makes them more susceptible to pests such as bark beetles, leading to the rapid destruction of vast forest areas. Some trees have developed defence mechanisms, such as expelling sap into the beetle-made holes before the insects have a chance to lay eggs. Moreover, intense heat and dry vegetation can lead to hotter, more widespread wildfires, making tree mortality more likely. In Rosenblad's study, the most dramatic changes in forest composition were observed in the plots located on north-facing slopes, those experiencing extreme warming and drought, and those areas prone to insect attacks. Some trees have adapted to these harsh conditions by developing thicker bark and frequently self-pruning, that is, shedding lower branches that are no longer growing well and might otherwise assist a fire in reaching the canopy. Surprisingly, some trees have evolved to utilise wildfires as an opportunity for new growth, either by sprouting from a stump post-fire or by dispersing seeds that germinate in the aftermath of a fire.

Senior author of the study David Ackerly said in the press release, "This research shows that tree species with low temperature tolerances are dying, and new trees are not growing fast enough to compensate for these changes. These trends are an early warning that changes in the forest are lagging behind the pace of climate change, which may make them more vulnerable to warmer and drier conditions in the future."

Solutions and debates

Environmental journalist Brian McHugh recently penned an article addressing the peril British trees face owing to climate change and potential solutions.[10] He pointed to a *Financial Times* article stating, "Aggressive agricultural and forestry practices, habitat loss and climate change are the main factors blamed. But while farming practices can be improved and habitats protected, does the pace of climate change mean we must accept the need to embrace non-native plants, specifically trees – that can cope with our changing weather? We only have about 40 native tree species in the UK, and many are struggling."[11] This suggests that non-native trees are more adaptable to climate change, and we might need them. However, whether to prioritise native or non-native trees to survive and even fight climate change is an ongoing debate.

McHugh mentioned a 20-year study conducted by the Botanical Society of Britain and Ireland, which discovered that over 50% of the native plants in Britain had declined owing to human impacts and that non-native plant species now outnumber native 'wild' species.[12] He mentions that botanical gardens are struggling to keep their trees in a healthy state even with a dedicated team of arborists to care for them. Many landscape architects are now experimenting with planting trees from slightly warmer climates. The aim is to find trees adapted to the rising temperatures that the UK (or other countries) will likely endure in the coming years owing to climate change. Trees have historically and naturally migrated in response to changing climates over long periods. However, the current climate change is happening much more rapidly, outpacing the natural migration abilities of many tree species. The concept of 'assisted migration' attempts to bridge this gap by facilitating the movement of trees to more suitable habitats before they are negatively impacted by changing conditions.[13] The potential benefits of assisted migration include helping tree species adapt to new environments and maintaining ecosystem function and biodiversity. By selecting trees from warmer regions that exhibit traits enabling them to thrive in anticipated future climates, restoration scientists and landscape architects hope to foster resilient ecosystems. However, debates and concerns surround the practice of assisted migration. Critics argue that it may lead to unintended consequences, such as the displacement or extinction of native species. Introducing non-native trees may disrupt existing ecological relationships, impact local biodiversity and reduce the adaptive capacity of native species. Balancing the potential benefits with the risks is a key challenge. Another debate centres around the uncertainty and complexity of climate change impacts. Predicting future climate patterns locally is challenging, and there's a risk of selecting tree species that may not be well suited for the actual conditions that emerge. Some argue that enhancing the resilience of existing ecosystems and supporting the migration of native species may be a more cautious and ecologically sound approach. Careful planning, research and monitoring are essential to address these concerns. Non-native plants often grow faster than native plants. This can be for various reasons; for instance, certain adaptive traits can allow them to outcompete native species. Moreover, native plants have co-evolved with other organisms in their native habitats, establishing complex ecological relationships. When introduced to new environments, non-native plants may not face the same co-evolved interactions, enabling them to allocate more resources towards growth rather than defence against pests, diseases or herbivores. However, non-native plants can also decompose faster.[14] This phenomenon has been shown to accelerate the release of 150% more CO_2 from the soil and into the atmosphere, thereby contributing to a potentially vicious climate cycle. Indeed, one study demonstrated that non-native plants provided a better food source for insect herbivores than native plants, which resulted in more plant-eating insects in communities dominated by non-native plants.[14] The authors used 160 experimental plant community plots and found that non-native plants also raised the abundance of soil microorganisms involved in the rapid decomposition of plant material. The intricate synergy between various organisms, involving fast-growing plants with less dense tissues, high insect herbivore abundance and increased decomposition by soil microorganisms, resulted in a higher proportion of plant carbon being released back into the atmosphere.

Many people are planting non-native trees to help solve one of the most profound issues of our era (climate change), but in some instances they might be exacerbating the very issue they aim to resolve. These are important points to consider in the climate change and forest restoration debate. Two things are abundantly clear: the task is urgent and complex. Restoring biodiverse forests (along with other habitats such as grasslands) can help reduce the impacts of climate change. If done globally, it could potentially slow and reverse global warming. But we must be cautious when planting non-native species willy-nilly. Careful planning and ecological thinking have never been more important. Moreover, as Brian said, "The dangers of monoculture plantation can be evident when disease strikes. We need to be cautious when politicians use tree-planting numbers as evidence of their 'green' credentials."

The albedo effect

There's also the intriguing concept of the albedo effect to consider. Albedo is the fraction of sunlight that's diffusely reflected by a body. It's measured on a scale from 0 (the full absorption of radiation) to 1 (the full reflection of radiation). I mentioned in Chapter 5 the Great Green Wall – the inspiring project that seeks to create a natural green belt across Africa. One of its core aims is to transform the arid and degraded land into a thriving forest ecosystem through mass tree-planting and restoration efforts. However, a recent study published in *Science* reveals an unexpected downside to such reforestation and afforestation endeavours.[15] It suggests that these projects could actually contribute to increased warming. Because of the albedo effect, dark-coloured forests, compared with light-coloured dry lands composed of shrubs and grass, absorb more radiation from the sun, leading to a potential rise in temperatures. While the study does not undermine the overall importance of growing and restoring forests to sequester carbon and enhance biodiversity, it does shed light on a significant drawback.

One of the coauthors, Dan Yakir, an ecophysiologist at the Weizmann Institute of Science in Rehovot, Israel, emphasises the need to consider the broader impact on the climate when discussing tree planting. He recently told the Proceedings of the National Academy of Science Journal Club: "When we talk about planting trees, we often focus on only reducing CO_2 when what we should care about is the overall impact on the climate."[16] The study aimed to balance these effects and revealed that, in many cases, tree planting could result in a net warming impact, serving as a wake-up call for further consideration. Owing to the albedo effect, the contrasting characteristics of bright desert lands, which reflect sunlight, and dark forests, which absorb it, play a significant role in climate regulation. This same principle contributes to the Arctic's warming as shiny sea ice melts, giving way to dark water. While these findings highlight potential challenges, they also underscore the importance of understanding the complex interactions between forests, land and climate. Ongoing research and a holistic approach will help refine strategies for sustainable and effective climate action, guiding future reforestation initiatives towards their intended goals.

Forest restoration is a critical endeavour that demands our attention, innovation and collaboration, especially in this rapidly changing climate – but it comes with underappreciated complexities. Informed action is key. Seek it out in all its manifestations. It can be a catalyst for change and, together with unwavering hope, empower us to shape a better future.

CHAPTER 17

Future Forests and Tomorrow's Guardians

"The greatest value of a picture is when it forces us to notice what we never expected to see."
—John W. Tukey

Future forests

The fate of forest ecosystems holds profound implications for planetary wellbeing. Our journey through this book has unravelled the threads of both hurdles and hope, and there will undoubtedly be myriad more challenges to navigate ahead. As stewards of the Earth and as a species that has an outsized impact on the environment, the choices we make today will reverberate through the canopies of tomorrow, shaping the destiny of forests for generations to come. But what are the possible future scenarios for forests? And how can we nurture the next generation of forest guardians?

Forestry

Vast stretches of land we call 'forests' are merely shadows of what forest ecosystems should be. For instance, most plantation forests were designed with one goal in mind: to produce timber. Consequently, many attributes of flourishing forest ecosystems – diversity, complexity and resilience – are lacking. We simplify and densify, and so we end up with forests that are very uniform in their age, size and structure. If the forestry industry is to continue, we must revisit the way we manage (or collaborate with – see Chapter 15) the forests. If we're truly committed to safeguarding nature, this entails abstaining from the practice of planting trees for short intervals, harvesting them, then replenishing them, and continuing the cycle. It means not using intensive herbicides and other harsh chemicals. Ultimately, it boils down to not treating forests like they're industrial farms for growing timber.

However, there's hope, and *ecological forestry* might be one of the keys to achieving a more harmonious balance between human needs and the preservation of forest ecosystems.[1] This approach recognises that we may need timber for our products, but we can harvest sustainably while leaving old-growth forests alone. In ecological forestry, you open some of the forests, allowing early succession and shade-intolerant species to thrive, but you retain significant areas of mature and younger trees and protect large areas of the ecosystem intact. In theory, this preserves the complex structures and biodiversity in the system. Ecological forestry also uses longer rotation periods, as much as 120 years between harvests.[2] This contrasts with plantations, which are typically 'simple' systems based on quick rotations.

Amidst its merits, ecological forestry is not without criticism or challenges. The principles of ecological forestry may be difficult to apply in highly fragmented landscapes or in areas with competing land uses, and this poses a considerable challenge. Management at the ecosystem scale is also emphasised, which may not always align with the interests or capabilities of individual landowners or managers. Achieving meaningful ecological outcomes may also require cooperation and coordination across multiple land parcels or jurisdictions, which can be logistically challenging. Furthermore, while this form of forestry could offer long-term ecological benefits, the upfront costs or reduced timber yields may prevent its uptake. Whatever happens, in my opinion the path forward for forestry lies in embracing a paradigm shift. We've harvested trees for millennia and this is unlikely to change; what demands our attention is how we harvest and how we shape the stewardship that must accompany it.

Climate change

We explored the effects of climate change on forests in Chapter 16, so I'll keep this relatively brief. There's no crystal ball to tell us how future forests will respond to climate change, but we can make predictions using the vast amount of data collected by scientists each year. An interesting study coming out of the University of Arizona combined tree-ring data with US Forest Service inventory data of

Arizona's ponderosa pine forests.[3] They used these data to enhance the forecasting of tree growth in future climate-driven scenarios. Shockingly, the researchers predicted a decline of 56–91% in individual tree growth over the period 2020 to 2100. The higher temperatures basically stunt the trees' growth. These declines in growth also mean Arizona's forests will likely take up less atmospheric carbon dioxide in the future. It's not clear in the study, but I believe these trees are mostly in plantations and, as I mentioned earlier, this already poses challenges. The authors suggest that the study has implications for forest managers who have an important opportunity to mitigate climate-related stress on the trees by changing the tree density to reduce competition from neighbours. If you have an overly dense forest and climate change happening at the same time, it's double the trouble. This only strengthens the case for ecological forestry and weakens the case for vast monocultures.

Not only will the geography of forests be reshaped in the future, but the immense natural forces of climate change will mould their physical attributes and reduce their capacity to store carbon – especially in monoculture plantations where simplicity stifles biodiversity. We can either turn a blind eye, shake the unseen hands of time and accept a fate of continued degradation, or use this forecasting to intervene and be the change-makers that our planet needs.

Fusion – a magic bullet?

What about future technologies that promise to revolutionise the energy industry and reduce our reliance on natural resources, thereby alleviating the pressure on global forests? Whether or not this promise is fulfilled is a different story, but let's take nuclear fusion for example. We've already spoken about nuclear *fission* (the splitting of atoms to generate energy) and its catastrophic risks back in Chapter 7, but nuclear *fusion*, on the other hand, is the fusing together of atoms to generate energy. It's inordinately more complicated than this, but that's the basic premise. Nuclear fusion is billed as being considerably safer and cleaner than its cousin, nuclear fission, and in theory it could provide near-endless energy.[4] As with a lot of things in life, many people love it while many other people hate the idea. Critics say it's expensive, that scalable fusion power remains elusive and that the reactor materials can become radioactive and dangerous.[5] They also reckon that focusing on fusion might divert attention and resources from existing, proven renewable energy sources like solar and wind. However, its proponents say that we've made major technological breakthroughs in recent times, it's considerably safer than fission, and harnessing existing renewable energy sources involves expensive and earth-damaging processes that aren't sustainable.

Let's jump into a time machine again, but this time we'll turn the dial to the future: the year 2100. If – and it might be a big IF! – proponents of nuclear fusion are correct, then we should have several functional nuclear fusion reactors on our planet by then. Now imagine that in 2100 we mostly use nuclear fusion energy, perhaps with some solar and wind on the side. We've considerably reduced our reliance on fossil fuels, which are major drivers of deforestation when extracted or harvested for energy production. Back in our own time (2024 at the time of writing), forests are cut down to provide biomass for cooking and heating in many parts of the world. If nuclear fusion or other clean energy sources become more widely available, there could be a reduced demand for biomass energy, thereby lessening the pressure on forests. Fusion energy might also find applications in industrial processes that currently rely on fossil fuels. If these processes transition to fusion-powered alternatives as we might see in the year 2100, it could decrease the demand for raw materials often sourced from forests for manufacturing and production.

Now, let's step outside of our time machine. What do we see? Vast flourishing forests and other ecosystems as far as the eyes can see...? This is a possible (and dare I say, ideal) scenario. It might go something like this: We have this supposed near-endless and 'clean' energy source, so we simply let the forest naturally regenerate and live in harmony with them. The energy is shared equally across the planet, and this reduces social inequity and pressures on food systems. We lightly manage our urban landscapes but don't dominate the natural world. We have strongly upheld international regulations that protect and restore ecosystems. This helps mitigate climate change and promote biodiversity. The air we breathe lacks pollution and is thronging with life-sustaining biochemical compounds and diverse microbiota. The diversity and integrity of forest ecosystems is something to behold.

Of course, life's never that easy and there are alternative, less rosy scenarios to consider. We might step outside of our time machine only to find degraded environments. Despite no longer relying on fossil fuels and needing to chop down trees for energy, we've now also got more affordable energy than we can shake a stick at (pun intended). It allows us to extract materials easier and quicker than ever before. It also lets us build giant structures and develop technology faster and faster. This is a dangerous self-reinforcing cycle. Humans, at least in some cultures, seem to take everything for granted – our luxuries, our relationships, our environments. Psychologists call the feeling of never being satisfied the 'hedonic adaptation' or the 'hedonic treadmill'.[6] For many of us, even if we get the job promotion we've been seeking, the newest and biggest car, or experience some other major change in our lives (and it can be negative too), our happiness peaks and then falls right back down to a baseline level. This means we're always wanting more – we're essentially adapting to this new state of affairs, and the initial pleasure diminishes. So, there's an argument that the combination of near-endless cheap energy and hedonic adaptation equates to a very risky scenario. We might fall into the hedonistic trap of extracting more elements, felling more forests and degrading more ecosystems to feed our infrastructural and technological greed.

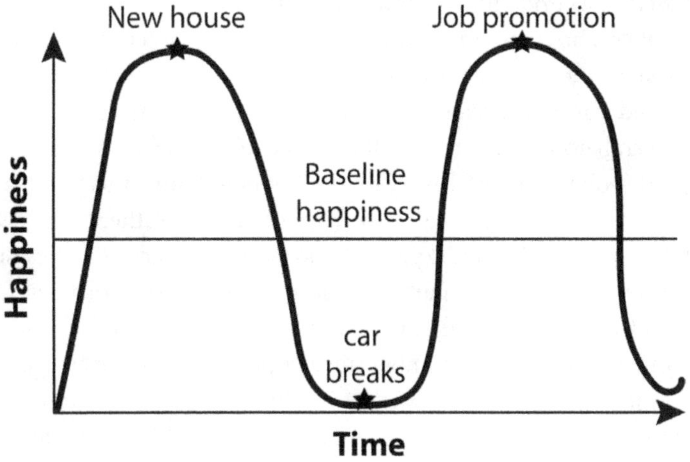

The hedonic adaptation or 'treadmill'.

There are many other possible scenarios – for instance, nuclear fusion energy might only have a neutral or minimal impact on global forests. I thought I'd take to social media to ask people's views on the matter. I deliberately omitted the word 'nuclear' because it can cause inflammatory reactions. Here are the results from a few dozen respondents:

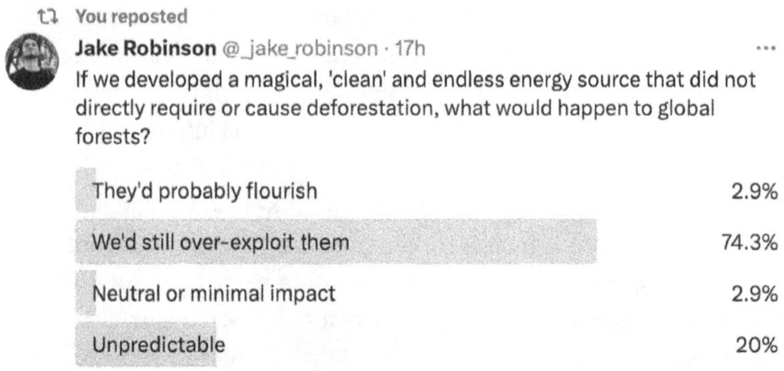

Results of my X poll.

It seems that most people (in my tiny study at least) think we would still over-exploit global forests. Only a few people thought forests would flourish if we no longer needed to exploit them for energy. Here's to hoping the minority prevail.

Urban forests and innovative architecture

We spoke about urban Miyawaki forests in Chapter 9, but there are some other, perhaps wackier, forms of 'forests' that have been proposed for our urban environments, including vertical forests. However, before I elaborate on that, let's take a quick look at floating trees or 'bobbing forests'. In 2016, the Bobbing Forest was presented as part of the National Tree Festival in Rotterdam, Netherlands.[7] Fear not, this wasn't a just quirky idea to boost green spaces in cities due to our tendency to 'pave paradise' and run out of space for trees. On the contrary, it was an art installation calling for greater awareness of climate change and the importance of protecting biodiverse green spaces in our cities. It consisted of 20 elm trees planted in modified maritime buoys. The buoys were packed with a blend of soil and ultralight volcanic rocks, aiding buoyancy despite their mass. This mixture helped to regulate both moisture and temperature levels. At the base of each buoy, a 500-litre tank was nestled, brimming with fresh water to nourish the trees. It was essential to replenish each water tank at least four times annually. The forest floated in Rotterdam's river for five years. I must admit, I was taken aback when I first saw these trees. It's quite a cool art project with an important message about the threats facing our future forests.

Bosco Verticale and friends

It's virtually inconceivable to discuss vertical forests without mentioning Bosco Verticale in Italy, which literally translates in English to the 'Vertical Forest'. The designers of Bosco Verticale state that it represents a pioneering model of architectural biodiversity, prioritising not only the wellbeing of humans but also fostering harmonious relationships between humanity and other living organisms. Situated in Milan's Porta Nuova district, this innovative structure comprises two towers, standing at heights of 80 and 112 metres respectively. Within its confines are (or were at the time of planting) a total of 800 trees – including 480 mature specimens and 300 smaller ones – alongside 15,000 perennials and ground-covering plants, as well as 5,000 shrubs. This vertical forest spans the equivalent of 30,000 square metres of trees and undergrowth, densely packed into a mere 3,000 square metres of urban surface.[8]

Bosco Verticale in Milan, Italy.

Bosco Verticale was first described as being a "home for trees that also houses humans and birds", and this is quite an accurate statement. One study suggested that years after its completion, Bosco Verticale provided habitat for various animal species, including more than 1,600 birds and butterflies.[9] Another study found that dense and green buildings in Milan, including Bosco Verticale, had a higher mean richness of bird species landing on them (=5) than similar large-scale buildings without vegetation (=0.83).[10]

It's safe to say that Bosco Verticale is one of the world's most iconic examples of 'green' architectural design. However, despite supporting hundreds of trees and birds, it's not without its critics. A colleague remarked that it was "an inspired act of architectural greenwash". Indeed, critics say the construction of buildings like the Bosco relies on vast amounts of carbon-intensive concrete. And yes, the trees in the balconies will absorb carbon. But it may take decades or centuries for them to absorb the amount of CO_2 that was emitted in the making of the balconies and planters. Another reason it's considered greenwash is because in order to live in the apartments you most likely need to be a multi-millionaire. Some Bosco apartments are purportedly worth upwards of £4,000,000. Thus, the building has a huge carbon footprint and was built only for the elite – essentially privatising nature at a high environmental cost.

Nevertheless, the designer of Bosco is now creating more affordable versions in other cities. For instance, in Eindhoven, Netherlands, there's the Trudo Vertical Forest which has 125 trees and 5,000 shrubs over 19 storeys, and tenants pay less than £800 a month to live there. So, the socioeconomic gap is closing.[11] Vertical forest buildings are popping up all over the world now, especially in China, and I'm guessing the carbon footprint of these buildings is also high. Yet the Bosco designer said they're now using carbon-friendly timber framing and shorter balconies that reduce the reliance on concrete.

However, using wood in construction may not be the climate-friendly solution it's touted to be. Researchers at the World Resources Institute found that using wood in construction probably increases carbon emissions relative to using concrete and steel.[12] They estimate a 54% increase in demand for wood by 2050. This would result in the clearance of woodland spanning an area equivalent to the size of the USA, totalling three million square miles. My view is that greening desolate urban surfaces and creating vertical forests has the potential to supplement broader urban greening initiatives, but there are some important caveats. The housing must be affordable for all, the environmental footprint can't outweigh the benefits of the vertical forests (from an ecological perspective, this is counterproductive), and they should not *replace* efforts to protect and restore diverse ecosystems on the ground. If these boxes can be ticked through social policies and innovative engineering, then vertical forests could play a role in sustainable cities of the future.

Return of the shifting baseline

My colleague Miles Richardson is a nature connectedness expert and is currently studying people's perception of and connection with trees.

In a recent blog, he observes that despite the UK being one of the most deforested nations on the planet, 80% of people in his study thought they had plenty of trees within a five-minute stroll of their home.[13] Around 25% of his participants also said there were plenty of woodlands within five minutes of their home. Miles's team found similar views on the presence of other natural features, such as wildflowers. These results perhaps reflect a 'shifting baseline' as discussed in Chapter 1, where each generation sees the current level of woodland as the norm. When participants were asked if they had a close relationship with trees, 16% felt this was very close. I actually thought this was quite a high proportion, especially considering that only 7% felt that they had a very close relationship with their neighbours! My former colleagues at the University of Sheffield found that some people love to spend time engaging with trees because "nature doesn't judge you".[14] Perhaps that goes some way to explaining the 7–16% gap.

Miles says asking about people's relationship with trees might seem unusual, but it's those who have a stronger bond with nature who take the most positive actions towards it. There's a nuance here that we should unpack. People might notice trees and have a close relationship with them, but shifting baseline syndrome may inhibit their drive to take meaningful action to protect them. As Miles remarks: "In some ways everything is a little too rosy as people tend to think there are plenty of trees and, therefore, unlikely to do a great deal to help restore woodland cover."

Augmented and virtual reality and nature connectedness

We know that technology isn't going away. So, could we use it to promote nature connectedness? A study recently demonstrated that exposure to nature in immersive virtual reality increases nature connectedness among people with a perceived low affinity for nature.[15] The participants were asked to answer questions from the Nature Relatedness 6 (NR-6) instrument, including "My connection to nature and the environment is a part of my spirituality", "I feel very connected to all living things and the earth" and "My relationship to nature is an important part of who I am", among others. They found that following the VR nature experience, nature connectedness scores incrementally increased and negative emotions decreased over three points in time. I have mixed feelings about this. Shouldn't we be focusing on promoting strategies that actually get people outside and interacting with the trees themselves? On the other hand, if this technology offers the impetus needed for people to start building their relationship with nature, perhaps it's a good thing.

Couldn't viewing nature in virtual reality potentially exacerbate the shifting baseline syndrome by providing individuals with a skewed perception of what constitutes a natural environment? As people become accustomed to experiencing nature through virtual simulations, they may gradually lower their expectations of what natural environments should look like in the real world or be tricked into assuming that the ecosystem is in perfectly good order and doesn't need to be protected or restored. I'd be interested to know whether the converse was true. Could we use VR to combat the shifting baseline syndrome (when people realise the pristine ecosystems they're viewing through a headset don't reflect reality) and promote positive action?

Ways to enhance our nature connectedness

Here are some simple tree-related activities to boost your nature connectedness:[16]

1. *Contact*
   Explore forests and trees. Take a closer look at tree bark and leaves with a magnifying glass. Engage with nature through all the senses – listen to the wind blowing through the canopy, smell the essential oils emitted by the trees, and caress the sometimes smooth and sometimes gnarly bark.
2. *Beauty*
   Take a few moments each day to appreciate the beauty of trees. Explore the fractal patterns of their branches, and engage in artistic activities – sketch a tree or make a collage from fallen tree parts.
3. *Meaning*
   Spend a moment each day to consider what trees and the rest of nature mean to you. Think about childhood memories in nature. Consider the ecological role of trees in providing habitat for thousands of animals and trillions of microbes. Think about how the spring blossom makes you feel.

4. *Emotion*
   Find happiness and wonder in trees and forests. Try to foster an emotional bond and love for trees. Reflect on your feelings and surround yourself with trees. Think about the size of trees, the intricacy of their evolution, and the memories they hold in their genes.
5. *Compassion*
   Consider what you could do for trees and the rest of nature. What meaningful actions could you take to protect trees? Extend your identity to include nature and nurture a more ethical lifestyle – one that's in tune with the wider biological community.

Trees and tots

When is the best time to sow and nourish the seed of nature connectedness in our minds to fortify our relationship with trees? Ideally, like many things, the sowing of this metaphorical seed begins when we're children. A study in Finland showed that trees satisfy many of the social needs of children.[17] They provide the materials, space and often purpose and contents for child play, including climbing, creative activities and hut building. Despite us becoming ever more risk averse when it comes to trees and children, the study suggested that kids learn to manage tree-related risks well and mainly from experience and playing with friends. Another study in Sweden suggested that a climbable tree turned out to be the main attraction in a playground, outcompeting the built play equipment, which was probably made of plastic and metal. Yet if trees are not available for children to play on and around, a deeper psychological connection between them is less likely to form.

Long-term interactions with trees and other plants during childhood can have a positive effect on appreciating trees and forests later in adulthood.[18] Playing in a place with trees can benefit child development and wellbeing, while continual contact can lead to a lifelong, personally meaningful sense of oneness with the natural world. To foster the next generation of forest guardians, we must allow children to interact with trees. To this end, my colleagues and I at Flinders University in South Australia are developing portable 'tiny forests' that can be installed on the grounds of daycare centres. The idea is to boost children's nature connectedness and their microbiome, while providing the versatility of moving the forest around.

Nature connectedness is not only positively associated with pro-nature behaviours but also egalitarianism.[19] So, what's the worst that could happen?!

I Once Was a Tree

From chewing gum, crayons and toothpaste to industrial chemicals, paper, petroleum and high-tech devices, trees shape our everyday lives in near-unimaginable ways. They have done so since our hominid ancestors first evolved and will continue to do so well into the future. It's also remarkable to consider that many of the atoms that make up the human body were once part of trees. The carbon, oxygen, hydrogen and other elements that form the building blocks of our bodies have cycled through the natural world for millions of years. Trees, with their ability to harness sunlight and convert CO_2 into organic matter through photosynthesis, play a crucial role in this cycle. As discussed in previous chapters, trees absorb CO_2 from the atmosphere and store it within their tissues when growing. Eventually, through various natural processes or human activities, these trees undergo decomposition, releasing their stored carbon back into the environment. This CO_2, can then be taken up by other trees or used by other organisms, sculpting our very existence. So, in an intriguing way, we can trace the origins of the atoms within our bodies back to the trees that once stood tall.

Take a moment to reflect on the multiple destinies of trees. They're in our vehicles, our books, our medicines, our breath, our blood, our bones.

CONCLUSION

Let Trees Be Thy Teachers

I began this book by asking, "What is a tree?" I conclude by saying, "A tree is a teacher."

If you're reading a printed copy of this book, you've sifted through 300+ compressed sheets of dehydrated pulp that were once, in another form, embraced by a hidden realm beneath our feet. Each page was part of an ecosystem in which cavernous filaments protruded and engaged in a mutual exchange with myriad subterranean life forms. The pages were once part of giant entities that reached skyward, their branches unfurled to reveal a sanctuary for creatures from birds to bugs. Each page was attached to leaves or needles that once 'breathed' in gases, and as they fell, they returned to the soil and enriched the earth. A cycle of death and rebirth. Their chlorophyll gleamed, converting sunlight into life through wondrous chemistry. Oxygen released, a gift to the masses.

Within each page lies a profound connection to nature, a reminder of the organic bonds linking all living things. As we 'leaf' through the bound remnants of forests, we hold in our hands a profound form of wisdom, reflecting the symbiotic and life-giving forces of the natural world. In the quiet solitude of reading, let's pay homage to these vessels of enlightenment. For not only do they teach and support us when they are alive, but they've also been the linchpin of science and culture, and continue to spread knowledge, language and inspiration in book form.

We've explored the evolution of forests – from the early pioneers that breached the primordial shallows of coastal waters and the ancient sentinels that survived several mass extinction events to the giants of our time covering a landmass of vast proportions. We know that humans have coexisted in tight-knit relationships with trees and forest ecosystems for millennia and continue to do so today. However, we also know that humans (some cultures more than others) have cleared billions of hectares of forests to feed our extravagant lifestyles while seemingly sleepwalking into an existential crisis. Chapter 2 on deforestation was admittedly challenging to write and perhaps to read, but it's a history that's part of our truth. The nuances of the narrative are important to acknowledge because the reality of the situation is often more complex than the media headlines will have us believe. Moreover, many people rely on harvesting trees for their livelihoods – and the real problem lies with the pattern of consumption in wealthy nations.

Speaking of people, we explored the social importance of trees, highlighting the cultural differences in how humans interact with them. Some people literally worship trees. Some treat them as shrines. Some remember and honour their dead relatives through intricate tree carvings. Some people get arrested for standing up for trees that are perceived to be nuisances in cities. This is how valuable they are to people. Trees are givers of life. And not just human life, but birds, insects, mammals, microbes and more. Forests are wildlife supporters, oxygen providers, soil protectors and carbon storers. Their ecological importance cannot be overstated. They feed moisture to giant flying rivers and influence – and are influenced by – planetary systems. Felling swathes of trees in one location can potentially shape the weather patterns in another location hundreds or even thousands of miles away – a teleconnection.

We've also considered the many faces and secrets of forest restoration. The number of people working to restore forest ecosystems around the world is inspiring. Innovative approaches, from using non-native nurse species to eavesdropping on the soil using ecoacoustics technology, all contribute to solving restoration puzzles. As physicist Edward de Bono said, "Lateral thinking is concerned not with playing with the existing pieces but with seeking to change those very pieces."[1] Lateral thinking will give rise to novel solutions in the face of extreme challenges.

But in some contexts, taking a step back and allowing nature to heal is probably the best option. Natural regeneration is a powerful force that runs through everything in life – from cells to bodies to whole ecosystems. Sometimes we need to give a helping hand when it comes to restoring forests, and we should always foster a culture of reciprocity with the land.

I must thank you all for enduring (but hopefully enjoying) my mildly obsessive excursion through the potential of oak tree and jay co-evolution. If nothing else, the role jays play in oak forest restoration is nothing short of extraordinary.

We explored the realm of radioactive restoration – again emphasising the notion that nature will find a way. We're yet to determine if the strange mutations in the Chernobyl trees are necessarily bad for their fitness. And the acoustic evidence for bird populations in the Red Forest, suggests that, at least in terms of numbers, birds are doing OK. However, more research is needed to see if their health and wellbeing are affected. Moreover, the risk of radioactive matter affecting other ecosystems and settlements outside Chernobyl remains. This is especially true when considering the increasing risk from climate change-associated wildfires. As writer Jane Braxton Little observed in *The Atlantic*: "Larger, more intense fires could destroy the forests entirely, obliterating their ability to keep what's in Chernobyl in Chernobyl."[2]

One of the main drivers of deforestation is unsustainable farming practices. However, alternatives are available, and organic and land-sharing practices might not be the way to go, especially as they pressure habitats in faraway places – out of sight, out of mind. However, as our German friend Klaus showed us, syntropic agroforestry might be a viable solution. Working with nature – using principles of strata, lifecycles and natural succession – to produce our food could be an overlooked forest restoration and conservation practice. It works in South America and Aotearoa, and Klaus is convinced it can work in temperate environments too. Hopefully, we'll all be living our own version of "tripping over Kiwi birds" in the future.

The Miyawaki method of creating 'tiny forests', particularly in urban areas, holds promise. I've certainly witnessed the biodiversity value. But the usual cautionary tales must be considered. Highly intensive tree planting could harm sensitive ecosystems, and frankly, we haven't done the research in all areas to know what the impacts might be. Still, I think it can, in some circumstances, be better than the alternative of a monoculture amenity grassland or a highly degraded city plot.

Amid the hope, awe and inspirational success stories, restoring forests can also be a controversial business. Mass tree planting can potentially cause more harm than good. This is especially the case when local ecological knowledge is not acquired (or is even ignored) or when people depend on the current land that has been earmarked for indiscriminate planting. In wealthy nations, we run the risk of big data imperialism by deciding on the areas that need to be restored without explicitly collaborating with the people who live on the land. Moreover, protecting the rights of people who live in a reciprocal relationship with forests is vital. And it's the patterns of wealthy consumers and the green washing that comes with them that we need to tackle.

To be truly restorative, whether ecologically or socially, we must take the time to learn about underappreciated perspectives. I think this is possibly the most important message to take away from this book. This is why I dedicated a whole section to it. Whether it's the invisible world (I often like to use the phrase "all the nature you *can* see intimately depends on all the nature you *can't* see"), the ecological wisdom of other cultures or the potential of looking 'beyond the brain' when it comes to cognition, learning about and embracing these phenomena is key to simplifying complexity and promoting transformational change. I think that another key or secret to such change is truth-seeking and moving beyond idle curiosity. Truth-seeking can manifest as proactively understanding the complexities of an ecosystem targeted for restoration. It could be understanding the culture of the people who ultimately steward the land. Truth-seeking could be understanding the psychological drivers behind our values and behaviours or teaching ourselves about the origins, transformation and destiny of trees. Truth-seeking can often be liberating, but it's always necessary. Let trees be thy teachers.

AFTERTHOUGHTS

Insights from Environmental Psychology

Recall the words of Aldo Leopold quoted in the Introduction: "There are two spiritual dangers in not owning a farm. One is the danger of supposing that breakfast comes from the grocery, and the other is that heat comes from the furnace."[1] The same concept applies to paper and other wood-based materials. As our bond with (the rest of) nature weakens, we must acknowledge the risk of presuming that timber and paper come from the shops, overlooking the intricate ecosystem that provides these resources. However, the fate of felled trees around the world extends beyond the wooden and paper items we commonly perceive. The transformation and destiny of these trees are far more complex, peculiar and astonishing than one might initially imagine. But how do these processes pass us by? How do we reconcile ourselves with our contribution to environmental harm? Are we even aware of our impact? And can the pressure of feeling helpless cause anxiety? The psychology underpinning our cognitive blindspots and our tendency to act in a manner that's either pro- or anti-environment is an intriguing topic that deserves some discussion.

The value–action gap

In the field of environmental psychology, there's a concept called the value–action gap. This is the disconnect between our values, beliefs and intentions and our actual behaviours and actions.[2] It highlights the disparity between what we say we value or intend to do and what we actually do in practice. This phenomenon is observed across various aspects of life, including personal ethics, social issues and environmental sustainability. Most of us are prone to falling down this gap at some point.

Sometimes we know that our actions fail to align with our values. It's like having a compass that points us in the direction we want to go, but we consciously walk in the wrong direction. We're aware that we're not living up to our own standards, but life's complexities get in the way. We can also get lost in a maze without realising it. We continue to make choices and behave in ways that contradict what we truly believe in, but we don't even notice the disconnect. It's also easier to spot this in other people rather than ourselves. This particular mental glitch is known as the bias blindspot.[3] It refers to the tendency for individuals to recognise cognitive biases and errors in others while being less aware of or less likely to acknowledge those same biases in themselves. Essentially, people tend to believe that they are less susceptible to biases and errors than others, leading to a blindspot in self-awareness regarding their own cognitive biases.

The value–action gap arises due to a multitude of factors. It can be influenced by cognitive dissonance, where individuals experience discomfort when their beliefs or values conflict with their behaviours. It can also stem from external factors, such as societal norms, convenience, lack of resources, or competing priorities, which make it challenging to translate intentions into concrete actions. The value–action gap is particularly evident in areas related to sustainability and environmental stewardship, which is directly relevant to forest restoration. For example, we often express concerns for the environment, support conservation efforts and advocate for sustainable practices, yet we fail to consistently act in accordance with these values.

It's not a black and white phenomenon, though, which means it's tricky to spot and do something about. Several years ago, I used to do ecology surveys in different parts of the UK. I care deeply about the environment and thought my job ticked the sustainability boxes. However, sometimes I'd drive all the way to Scotland from Derbyshire in a petrol-powered vehicle to essentially look at a field and jot down some notes. For me, this could be viewed as a value–action gap. I believed I was doing a good thing for nature by assessing the impact of windfarm developments. However, my journey involved spewing greenhouse gases into the atmosphere, killing hundreds of invertebrates (cars wipe out insects, including many pollinators) and releasing toxic particles from tyre wear into the environment. These negative side-effects were wrestled to the sidelines in my mind, probably driven by my focus on the positive effects of the job itself rather than the impacts of the whole process (the job plus the journey). But I think that the need to get paid for the work and the standards of practice (the 'norms') in the industry also influenced this blindspot. Turning down the work might have been the ethical thing to do. But it would have risked causing tension at work and potentially losing my job – a risk that seemed amplified as a junior back in the day. It's not always easy.

The gap between stated intentions and actual behaviour poses a significant challenge in addressing pressing environmental issues. It also often occurs at the group or organisational level. For instance, organisations or governments may endorse policies or commitments to promote forest restoration but fail to implement concrete actions or allocate sufficient funding and resources to support these initiatives effectively. This, of course, can also be steeped in manipulation and dishonesty, which in some cases seems to be a prerequisite for a career in politics – the very discipline we need on board to drive positive change.

Closing the value–action gap requires awareness, reflection and conscious effort. It entails consciously aligning our values and intentions with our day-to-day actions while not feeling too guilty when we fail to do so – after all, as humans, we're prone to making mistakes. Trying to do what we can is key. This conscious alignment can involve making small (but long-term) changes in our habits and adopting sustainable behaviours. For those in power, it also necessitates creating supportive environments that make sustainable choices more accessible and convenient for individuals. Bridging the value–action gap is a challenging task, but it's crucial for effecting meaningful change.

However, the burden of environmental damage should not be placed solely on individuals. Environmental damage is often the result of systemic problems, such as unsustainable industrial practices, inadequate regulations and lack of corporate responsibility. Placing the blame solely on individuals ignores the larger structural issues contributing to environmental harm.

Eco-anxiety

"I'm afraid of the future," said Clarissa, my friend's (very smart) ten-year-old. "Why?" I asked. "Because the planet is burning and it's going to get worse." This is a lot for a child to process and worry about. However, the anxiety is real and rising. A recent study of 10,000 young people (aged 16–25 years) found that 59% were very or extremely worried about climate change, with 84% at least moderately worried.[4]

'Eco-anxiety' is a term used to describe the psychological distress, worry or fear experienced by individuals in response to the environmental challenges facing our planet.[5] It arises from concerns about climate change, biodiversity loss, deforestation, pollution and other ecological issues. As people become more aware of the urgency and magnitude of these problems, they can feel overwhelmed, helpless or deeply saddened by the state of the environment and the potential consequences for future generations.

Forest degradation can play a significant role in the onset of eco-anxiety. But on the flipside, forest restoration can play a part in addressing it. Engaging in forest restoration efforts can empower individuals to take tangible action and contribute to positive change. By actively participating in tree-planting (caring) initiatives and conservation activities, individuals can find a sense of purpose and hope, knowing that they're actively working towards restoring and preserving vital ecosystems. I've worked alongside restoration volunteers for years and can attest to the profound impact of direct action on individuals' lives. It often enhances wellbeing, infuses lives with purpose and cultivates a sense of giving back to the community and the ecosystem.

Dealing with eco-anxiety requires a holistic approach that combines individual coping strategies and collective action. Self-care and mindfulness are important for overcoming eco-anxiety.[6] Engaging in activities such as meditation, spending time in nature, exercise or creative outlets that promote relaxation and a sense of peace can all help. I'd like to acknowledge here that social inequity issues can make these activities more challenging. 'Time poverty', for instance, owing to working more than one full-time job to pay the bills can reduce a person's available time to engage in meaningful activities – it's often a systemic issue that needs government intervention.[7]

Seeking support and connection is also important. Share concerns with friends, family or support groups who understand and empathise. Connect with likeminded individuals or join environmental organisations that actively work towards the restoration of nature. Collaborating with others can offer a sense of community and collective strength. Our collective environmental efforts can yield significantly more impactful outcomes than the sum of our individual actions. Taking direct action in protecting forests can instil a sense of purpose and empowerment, and using our voices to raise awareness about environmental issues can be empowering.

While eco-anxiety is real and understandable, channelling these emotions into positive action is essential. By caring for ourselves, connecting with others and actively participating in restoration efforts, we can make a meaningful difference and find solace in contributing to the planet's healing.

Learned helplessness

A recent study published in the *Journal of Environmental Psychology* investigated the role of learned helplessness in environmental action.[8] This is a psychological condition where an individual has learned to behave helplessly after repeatedly feeling as if they have no control over a situation. The researchers found that learned helplessness was a barrier to pro-environmental behaviours such as recycling and conservation. Even if someone is concerned about the environment, they're less likely to take action if they're experiencing learned helplessness. This finding is significant because the scope of the environmental problems we're currently facing is so vast that individual behaviours can feel almost meaningless. The activities already mentioned – from self-care to joining forest restoration groups – can also help reduce helplessness.

Pro-environmental behaviours

A recent study suggested that pro-environmental behaviours can be underpinned by various psychological theories, including the 'theory of planned behaviour'.[9] This concept helps us understand how our decisions translate into actions. Imagine you're considering starting a new fitness routine. According to the theory of planned behaviour, your decision to hit the gym isn't as simple as just deciding to go. Instead, it's a three-step mental journey that starts with your attitude towards exercise. You'll think about the likely outcomes: stronger muscles, improved mood and better health. If these outcomes sound appealing, you'll have a positive attitude towards working out. But your personal beliefs are just one part of the equation. Next, you'll consider the societal expectations or 'subjective norms'. This might involve thinking about how your friends, family or co-workers view exercise. If they're all fitness enthusiasts who love talking about their workout routines, you might feel more social pressure to start exercising. The final piece of the puzzle is 'perceived behavioural control', which is essentially your belief in your ability to carry out the behaviour. Even if you have a positive attitude about exercise and feel that your friends expect you to do it, you might still not hit the gym if you don't believe you have the time, resources or willpower to do so.

By understanding this interplay of attitudes, societal expectations and personal control, the theory of planned behaviour provides a framework for predicting how our intentions become actions. Therefore, when you find yourself mulling over a decision, remember that you're probably embarking on a three-step psychological journey. Studies have shown that researchers can predict the impact of these attitudes, subjective norms and perceived behavioural control on people's intentions to engage in pro-environmental behaviour (such as protecting forests).

My final example of a theory to explain pro-environmental behaviours is the 'norm activation theory'.[10] Imagine walking through an urban forest and seeing a piece of rubbish on the ground. You look around and notice a rubbish bin just a few steps away. What do you do? Norm activation is a psychological theory that helps us understand why some people might pick up the rubbish while others walk past it. The decision to pick up the litter starts with the recognition of a problem – the urban forest, a place for everyone (including other species!) to enjoy, is being spoiled by litter. This realisation awakens a personal norm, an internal rule that guides our behaviour. In this case, the norm might be a belief in keeping shared spaces clean for everyone. But, as the theory suggests, this norm doesn't spring into action automatically. Two key conditions need to be met.

First, you need to feel a sense of responsibility. You didn't throw the piece of rubbish on the ground, but if you believe keeping the forest clean is everyone's duty, you'll feel personally responsible for addressing the issue. This is called 'ascription of responsibility'. Second, you need to understand the consequences of your actions or inaction. This is called 'awareness of consequences'. You're more likely to act if you realise that not picking up the rubbish could harm the environment, spoil the forest for others or encourage more littering. So remember to think about the norm activation theory when pondering decisions that affect the environment. It shows us that our actions are not just about personal preferences or needs but about how we perceive our responsibilities and the impact of our actions on the world around us. While the connection isn't always straightforward, I believe that seeking, recognising and accepting the truth frequently serves as a stepping stone towards experiencing a sense of responsibility and an obligation to take action. Again, the onus shouldn't be solely on the individual but, in fact, far more so on powerful institutions - and we need to band together and call them out at every opportunity.

In this exploration of the psychological frameworks influencing forest restoration, we find that the synergy of empathy, intrinsic connection to nature and a collective environmental consciousness forms the bedrock of success. As individuals cultivate a profound understanding of the interplay between the human mind and the natural world, the potential for effective restoration grows. Embracing the psychological dimensions embedded in forest restoration efforts not only nourishes the earth but also nurtures the mind.

Notes

Introduction: Seeing the Woods for the Wood

1. *A Sand County Almanac* is a 1949 non-fiction book by American ecologist, forester and environmentalist Aldo Leopold. This book inspired me on my journey into ecology.
2. In 1390 Ulman Stromer created founded the Gleismühle, Germany's first paper mill. Paper was often made out of rags.
3. Carl Sagan's famous quote from his TV series, *Cosmos A Personal Voyage.*
4. Margot Neale explains this well in Pascoe and Gammage (2021).
5. Trees can sense their surroundings and retain a memory of sorts (via epigenetic means) – this is the focus of the MEMBRA project: https://membra.info/.
6. Professor Suzanne Simard, in particular, is a strong proponent of the notion that trees can communicate with each other via chemical processes. However, the 'wood wide web' concept is hotly contested, as I describe in Chapter 13.
7. The vagus nerve is the largest cranial nerve in the human body and is thought to play a role in the parasympathetic nervous system and the microbiota–gut–brain axis. Acetylcholine, originally called vagusstoff, is released after stimulation of the vagus nerve. Acetylcholine was the first neurotransmitter to be discovered.
8. The UK has been named one of the most nature-depleted countries in the world. https://www.nhm.ac.uk/discover/news/2020/september/uk-has-led-the-world-in-destroying-the-natural-environment.html. According to the RSPB, more than 40 million birds have disappeared from the UK's skies since 1970.
9. According to Forest Research, the UK has 13% forest cover: https://www.forestresearch.gov.uk/tools-and-resources/statistics/statistics-by-topic/woodland-statistics/
10. The UK is lagging behind on forest cover compared with the broader EU region (which has 37.7%): https://www.europarl.europa.eu/factsheets/en/sheet/105/the-european-union-and-forests
11. Johnson et al. (2022).
12. Tropical forest loss in 2020: https://www.globalforestwatch.org/blog/data-and-research/global-tree-cover-loss-data-2020/
13. Yu et al. (2020). 95% of the land will be degraded by 2050 without immediate intervention. This has inspired movements such as the UN Decade on Ecosystem Restoration (2021–30). You may also see the associated hashtag #generationrestoration on social media.
14. Uexküll (1909). Yong (2023) talks about the Umwelt in detail.

15. This quote is from an address by Martin Luther King, Jr. in Washington, DC, in February 1968, two months before he was assassinated in Memphis.

Chapter 1: A Brief History of Forests

1. An explosion of new genes occurred as land plants emerged. This *Conversation* article provides a summary of the research (but also check out the academic papers within): https://theconversation.com/we-found-the-genes-that-allowed-plants-to-colonise-land-500-million-years-ago-128977.
2. See note 1.
3. Krám et al. (2009).
4. Rival et al. (2011).
5. This is a great *Conversation* article written by academic specialists on tree ferns with a mention of their importance in forming fossil fuels: https://theconversation.com/tree-ferns-are-older-than-dinosaurs-and-thats-not-even-the-most-interesting-thing-about-them-138435. See also Blair et al. 2017.
6. The Archaeopteris were spore-forming trees, sharing the most characteristics with the seed plants. Meyer-Berthaud et al. (1999).
7. DiMichele (2001).
8. Bowes (2010).
9. Smith and Botha (2005). Some estimate that up to 90% of marine species disappeared.
10. The earliest known, plant-eating dinosaur was the *Pisanosaurus mertii*, the fossils of which were discovered in the Ischigualasto formation of Argentina, South America: https://ebrary.net/3910/history/triassic_period_dinosaurs_herbivores. See also Agnolin et al. 2018.
11. De Min et al. (2020).
12. Whitlock (2011).
13. See note 12.
14. Hallam (1982).
15. Raven and Wackernagel (2020).
16. Friis et al. (2019).
17. The K-T event is thought to have caused a lingering effect on the environment for many of the years that followed, making it difficult for plants to photosynthesise.
18. Bralower (2008).
19. Davis (1983).
20. Poore et al. (2000).
21. Great *Conversation* article written by academic specialists on the last ice age: https://theconversation.com/the-last-ice-age-tells-us-why-we-need-to-care-about-a-2-change-in-temperature-126923#:~:text=As%20water%20turned%20to%20ice,would%20have%20found%20themselves%20inland. See also Anderson et al. 2019.
22. Pearce et al. (2023).
23. Rackham (1986).
24. Soga and Gaston (2018).
25. Mitchell (2005). This research looks at testing the Vera hypothesis.
26. Gammage (2013), also see Pascoe (2014).

27. Piperno et al. (2021) – the study suggesting Indigenous Peoples caused no detectable species loss in the Amazon.
28. Based on an analysis done by the World Resource Institute (WRI), part of the Global Forest Watch. Media article here: https://www.theguardian.com/environment/2020/jun/02/football-pitch-area-tropical-rainforest-lost-every-six-seconds.
29. Duarte et al. (2022).

Chapter 2: A Brief History of Deforestation

1. Our World in Data on deforestation: https://ourworldindata.org/deforestation#:~:text=10%2C000%20years%20ago%2057%25%20of,size%20of%20the%20United%20States. See also Ritchie and Roser (2021).
2. Armenteras et al. (2021).
3. Hebelstrup et al. (2023).
4. Pearsall (2008).
5. Wang et al. (2014).
6. Bell et al. (2022).
7. Richards (2011).
8. Fletcher et al. (2021).
9. Fuller (2011).
10. He et al. (2022).
11. Špulerová et al. (2022).
12. Forzieri et al. (2021). Forest cover will likely change year on year owing to timber production, wildfires (which are growing in intensity/frequency) and other sociopolitical and ecological factors.
13. Knudson et al. (2009).
14. Beresford-Jones et al. (2009).
15. Malhi (2018).
16. Rull (2019).
17. McAnany and Yoffee (2009).
18. Rull (2020). Archaeological evidence shows that the Rapanui civilisation was resilient to deforestation and remained healthy until European contact.
19. Cohen (2003).
20. A Royal Navy article talks about the 5,500 oak trees used to build HMS *Victory*. This article also talks about how Scottish estates felled 11 elm and 10 oak trees, donating them to HMS *Victory* refurbishing (as part of a 15-year conservation project). https://www.royalnavy.mod.uk/news-and-latest-activity/news/2016/february/15/160215-scottish-timber-hms-victory.
21. Maps show that nearly all North America's virgin forests have been removed. The maps were produced by the University of Michigan and hosted on this website: https://blog.richmond.edu/livesofmaps/2022/04/18/map-of-the-week-american-deforestation-over-the-years/.
22. The population of the USA per square metre of land in 1800 was 6.1, rising to 21.5 in 1900: a rapid population boom. This data was provided by the United States Census Bureau: https://www.census.gov/.
23. https://ourworldindata.org/land-use shows aggregated data and visualisations for global land use change. The original source of the data was Klein et al. (2016), using the HYDE 3.2 system.

24. According to Zhou et al. (2015), global urban land accounts for only 0.5% of the land cover. This varies between regions, from 0.1% in Oceania to 2.3% in Europe.
25. Deforestation data was acquired from https://ourworldindata.org/deforestation – based on data from Williams (2003) and UN FAO Global Forest Resources Assessment (2020).
26. See note 25.
27. See note 25.
28. The 260 million ha figure is based on a report by Climate Focus, suggesting 26 million ha of deforestation per year was reached in the last decade. Climate Focus (2019).
29. Crowther et al. (2015).
30. This calculation is based on an average tree height of 10 m. However, this is potentially a conservative estimate when considering the average height of a Sitka spruce tree used for construction in the UK is 16 m – 23 m when felled (Ramage et al. 2017). But we also need to take into consideration that not all trees are felled directly for the timber industry (especially when land is cleared for agriculture). The average distance to the moon from Earth is 382,500 km. 15 billion trees x 10 m = 150 billion m. If we divide this by 1,000, this gives us 150 million km. If we divide this by the distance to the moon, we get 392.1.
31. It's difficult to get an accurate figure regarding how many trees are planted each year. However, you will often find 1–5 billion quoted. A relevant *Science* article can be found here: https://www.science.org/content/article/earth-home-3-trillion-trees-half-many-when-human-civilization-arose.
32. Ferrante et al. (2019).
33. WRI (2023).
34. Pearce (2023).
35. See note 32.
36. Wells et al. (2022).
37. Kings (2022).
38. See Introduction Note 1 (Leopold, 1949).
39. Graziosi et al. (2020).
40. Tollefson (2020).
41. Robinson et al. (2022a).
42. Chua et al. (2003).
43. Schmidt and Ostfeld (2001).
44. Dempsey (2023).
45. Long et al. (2021).
46. Human-induced wildfires: https://www.nps.gov/articles/wildfire-causes-and-evaluation.htm.

Chapter 3: The Social Importance of Trees

1. According to Berry (1999), Japan's car makers had 21% of the market in 1980.
2. Pempel (1987).
3. Li et al. (2022).

4. Li et al. (2018).
5. Li et al. (2009).
6. Woo et al. (2019).
7. Robinson et al. (2021). You can find more information on the environment–microbiome–health axis on my website: www.jakemrobinson.com.
8. Šantl-Temkiv et al. (2018).
9. Rook (2012).
10. Robinson et al. (2021d)
11. Robinson et al. (2022d).
12. See note 9.
13. Robinson et al. (2020).
14. The James Hutton Institute: https://www.hutton.ac.uk/sites/default/files/files/Soils-A5-booklet.pdf. See also Timmis and Ramos (2021).
15. Robinson (2023).
16. Robinson and Breed (2019).
17. Robinson et al. (2020).
18. Lee et al. (2023).
19. Rahman et al. (2020).
20. Mavrogianni et al. (2022).
21. Robinson et al. (2022b) – see Chapter 2 Note 39. Increasing tree cover in urban centres is not only important for improving human health (e.g., via aesthetics, biogenic compounds, recreation value, etc., and reducing the urban heat island effect), but it can also provide wildlife corridors, connecting habitats so that animals can disperse across the landscape.
22. https://burialsandbeyond.com/2019/08/31/the-yew-tree-of-the-dead/.
23. See note 21.
24. Bjurhager et al. (2013).
25. https://ati.woodlandtrust.org.uk/what-we-record-and-why/what-we-record/oldest-trees-in-uk.
26. Buhrich and Murison (2020).
27. https://filmfreeway.com/Gaaguuwiyadhawungu-BringBackHome.
28. See note 21.
29. www.bbc.com/news/uk-england-south-yorkshire-35542110.
30. Woodstra and Allen (2022).
31. Das and Vincent (2019).

Chapter 4: The Ecological Importance of Trees

1. There are thought to be around 10,000 species of birds on the planet, and 50% of these live in the tropics (Lees et al., 2022).
2. World-renowned mycologist Paul Stamets once said, “I calculated that every footstep I take impacts more than 300 miles of mycelium.”
3. Stamets (2005).
4. Kimmerer (2021).
5. The Woodland Trust: https://www.woodlandtrust.org.uk/media/1833/wildlife-of-ancient-trees.pdf.
6. Mitchell et al. (2022).
7. See note 5.

8. Santos et al. (2021).
9. Tanacredi et al. (2019).
10. Harada et al. (1997).
11. Patino and Vanderpoorten (2018).
12. Elbert et al. (2012).
13. See note 12.
14. Armaleo et al. (2019).
15. Extreme plants thrive at 72°C in New Zealand hot volcanic soil: https://www.newscientist.com/article/2134206-extreme-plants-thrive-at-72c-in-new-zealands-hot-volcanic-soil/ and https://www.kew.org/read-and-watch/moss#:~:text=When%20they're%20desiccated%20(dried,resilient%20little%20plant%20right%20there.
16. Lydia et al. (2023).
17. Scientists recently used culture-dependent and independent (e.g., DNA sequencing) to estimate the total number of fungal species on the planet. This estimate was 12 million compared with 2.2–3.8 million estimated previously. Wu et al. (2019).
18. Shim (2022).
19. Antonelli et al. (2020).
20. Benner (2020).
21. https://ourworldindata.org/wild-mammals-birds-biomass.
22. See note 21.
23. Field et al. (1998).
24. Dai et al. (2022).
25. https://www.wri.org/insights/forests-absorb-twice-much-carbon-they-emit-each-year.
26. Meier et al. (2021).
27. Barlow (2016).
28. https://www.newscientist.com/article/mg24432540-600-earths-most-important-rivers-are-in-the-sky-and-theyre-drying-up/.
29. Aragão (2012).
30. Salati et al. (1983).
31. Van der Ent and Savenije (2010).
32. https://www.science.org/content/article/controversial-russian-theory-claims-forests-don-t-just-make-rain-they-make-wind
33. https://www.bbc.com/news/world-asia-india-18765284.
34. Bradshaw et al. (2007).
35. Steffen et al. (2020).

Chapter 5: Forest Restoration

1. Schultz et al. (2012).
2. https://www.emergenresearch.com/industry-report/artificial-grass-market.
3. Baker (1925).
4. Hanley (1983).
5. Men of the Trees (1933). Also check out Woudstra and Allen (2022). Camilla did her PhD on Richard St Barbe Baker.

6. Turner et al. (2021).
7. https://www.internationaltreefoundation.org/. Men of Trees later morphed into the International Tree Foundation.
8. https://news.mongabay.com/2022/03/indigenous-communities-transform-a-mexican-desert-landscape-into-forest/. See also Ramírez-Toro et al. (2017).
9. https://regenerationinternational.org/2022/04/24/indigenous-communities-transform-a-mexican-desert-landscape-into-forest/. See also Robson (2007).
10. https://www.wwf.fr/sites/default/files/doc-2020-01/202001_Report_10_years_of_restoration_of_tropical_forests_the_east_usambara_tanzania.pdf. See also Ndangalasi et al. (2021).
11. Bradby and Keesing (2016).
12. https://www.regennarration.com/episodes/82-geoff-bastyan#:~:text=If%20you%20haven't%20heard,lies%20the%20Oyster%20Harbour%20catchment.
13. Gagliano (2013).
14. Hussain et al. (2023).
15. Anthony et al. (2023).
16. https://www.wired.com/2008/10/darwin-earthwor/.
17. Robinson et al. (2021c).
18. Robinson et al. (2023).
19. Bjøgaas (2023).
20. Robinson et al. (2022c).
21. Kanchan et al. (2023).
22. Robinson et al. (2022c).

Chapter 6: Natural Regeneration

1. Yokouchi et al. (2016).
2. Margulis and Sagan (2005).
3. Crouzeilles et al. (2017).
4. Cook-Patton et al. (2020).
5. Crampton et al. (1998).
6. https://theconversation.com/monks-wood-wilderness-60-years-ago-scientists-let-a-farm-field-rewild-heres-what-happened-163406.
7. Kurek et al. (2019).
8. Dawkins (2016).
9. Dawkins (1976).
10. Wilson-Henjum et al. (2019).
11. Kremer and Hipp (2020).
12. https://www.newscientist.com/article/dn20035-lost-islands-of-the-crows-revealed-in-dna-study/.
13. Tree (2018).
14. https://www.cam.ac.uk/research/news/just-like-humans-more-intelligent-jays-have-greater-self-control.
15. Watts et al. (2018).
16. Selden and Putz (2022).

17. https://cosmosmagazine.com/nature/pocket-gopher-farm-roots/#:~:text=%E2%80%9CIf%20farming%20requires%20that%20crops,forests%20%E2%80%94%20but%20not%20necessarily%20planted.
18. Roberts and Todd (2017).
19. Evans (2010).
20. Tafoya et al. (2020).
21. Avalos et al. (2021).
22. Kappelle (2016).
23. Robinson (2023).

Paper – I Once Was a Tree

1. Tsien (1973).
2. Suleri (2018).
3. Del Rio et al. (2022).
4. Kumar (2017).

Chapter 7: Chernobyl's Red Forest

1. Rashad and Hammad (2000).
2. Beresford et al. (2023).
3. Scheidt (2018).
4. Kashparov et al. (2012).
5. Evangeliou et al. (2014).
6. https://www.theatlantic.com/science/archive/2020/08/chernobyl-fires/615067/.
7. Yao et al. (2014).
8. Ager (2019).
9. Lehmann et al. (2016).
10. Raines et al. (2020).
11. Møller et al. (2011).
12. The following article covers some of the critique that has gone the way of Møller and Mousseau: https://knowablemagazine.org/article/food-environment/2022/scientists-cant-agree-about-chernobyls-impact-wildlife.

Chapter 8: Agroforestry

1. Check out Permadynamics here: https://www.permadynamics.co.nz/ Klaus's family run outstanding educational courses that focus on establishing, researching, managing and teaching regenerative living systems to create and maintain community resilience and human-inclusive ecosystems.1
2. https://www.agroforestry.co.uk/.
3. Atangana et al. (2014).
4. Torralba et al. (2016).
5. https://www.agroforestry.co.uk/about-agroforestry/.
6. https://climate.org/.
7. Call et al. (2006).
8. Monbiot (2022).
9. Chavan et al. (2015).

10. https://www.downtoearth.org.in/news/india-becomes-first-country-to-adopt-an-agroforestry-policy-43518.
11. Chu et al. (2014).
12. Pe'Er et al. (2019).
13. Bateman and Balmford (2023).
14. Smith et al. (2019).
15. See note 13.

Chapter 9: The Miyawaki Method

1. Lewis et al. (2022).
2. Prach et al. (2016).
3. The Japanese tradition of chinju no mori has parallels with the concept of the sacred tree groves mentioned in Chapter 3. It's a core concept of the 'Shinto environmentalist paradigm' and represents an example of shrine-based conservation. Rots et al. (2015).
4. Singh and Saini (2019).
5. Miyawaki (1993).
6. Cárdenas et al. (2022).
7. Poddar (2021).
8. Teo et al. (2021).
9. https://www.orchardofflavours.com/miyawaki-inspired-food-forest.
10. https://science.thewire.in/society/urban/why-the-miyawaki-method-is-not-a-good-way-to-afforest-chennai/.
11. https://india.mongabay.com/2020/09/commentary-of-ants-and-antelopes-the-ecological-history-of-chennai-was-not-what-one-would-imagine/.
12. https://edepot.wur.nl/446911.
13. https://www.theguardian.com/environment/2020/jun/13/fast-growing-mini-forests-spring-up-in-europe-to-aid-climate
14. https://elcn.eu/on-the-ground/ivn-natuur-educatie.
15. Loidi and Fernández-González (2012).
16. De Nascimento et al. (2020).

Chapter 10: The Trillion Trees Controversy

1. Maathai (2003).
2. Mda (2016).
3. Lippke et al. (2021).
4. https://uplink.weforum.org/uplink/s/uplink-contribution/a012o00001pUEIDAA4/plant-a-million-zambia.
5. Bastin et al. (2019).
6. This article covers the lab leaders perspective: https://www.nature.com/articles/d41586-019-02846-4.
7. Veldman et al. (2019).
8. https://www.theatlantic.com/science/archive/2022/07/climate-change-tree-planting-preserve-grass-grasslands/670583/.
9. Fleischman et al. (2020).

10. https://www.worldagroforestry.org/publication/tree-planting-tree-growing-rethinking-ecosystem-restoration-through-trees.
11. https://www.worldagroforestry.org/blog/2020/02/17/planting-trees-or-growing-them-do-words-matter-restoring-land.
12. https://www.zeit.de/zustimmung?url=https%3A%2F%2Fwww.zeit.de%2Findex.
13. The UK's Department for Environment, Food and Rural Affairs, and confirmed by Friends of the Earth analysis: https://policy.friendsoftheearth.uk/print/pdf/node/151.
14. https://www.theguardian.com/society/2019/dec/28/replanting-britain-its-about-the-right-tree-in-the-right-place.

Chapter 11: Greenwashing Galore

1. Watson (2016).
2. Cherry and Sneirson (2012).
3. Rahman (2022).
4. https://www.forbes.com/sites/gregpetro/2020/01/31/sustainable-retail-how-gen-z-is-leading-the-pack/?sh=63733a0a2ca3.
5. Lanthorn (2013).
6. https://www.bbc.com/news/business-60481080.
7. https://www.theguardian.com/environment/2022/feb/02/activists-accuse-drinks-firm-innocent-greenwashing-plastics-rebellion-advertising-tv-advert?utm_source=dlvr.it&utm_medium=twitter.
8. https://www.theguardian.com/sustainable-business/2016/aug/20/greenwashing-environmentalism-lies-companies.
9. https://www.gov.uk/government/news/global-sweep-finds-40-of-firms-green-claims-could-be-misleading.
10. Sunder (1993).
11. Carse (2006).
12. https://www.oxfam.org/en/press-releases/net-zero-carbon-targets-are-dangerous-distractions-priority-cutting-emissions-says#:~:text=Using%20land%20alone%20to%20remove,a%20new%20Oxfam%20report%20today.
13. https://www.independent.co.uk/climate-change/news/cop26-climate-summit-indigenous-offsetting-b1951289.html
14. Clifford and Castonguay (2022).
15. Hoang and Kanemoto (2021).
16. Irwin et al. (2022).
17. https://www.sydney.edu.au/news-opinion/news/2022/04/13/your-morning-coffee-could-hasten-species-extinction.html
18. Albright et al. (2000).
19. Kopfman et al. (1998).

Chapter 12: Sea of Tree Guards

1. Chau et al. (2021).
2. Ritchie and Roser (2018).

3. Susanti et al. (2020).
4. Jiang et al. (2020).
5. De Souza Machado et al. (2019).
6. https://www.woodlandtrust.org.uk/media/47692/emergency-tree-plan.pdf.
7. Delate et al. (2021).
8. Beans (2023).
9. Szepesi (2021).
10. Ahirwar and Behera (2022).
11. https://www.hrw.org/news/2011/10/03/national-cashew-day-more-nuts.
12. https://impactpolicies.org/news/69/Cashews-are-Delicious-but-Come-with-a-Human-Cost#:~:text=Many%20of%20Vietnam's%20cashews%20are,abuse%20and%20extreme%20working%20hours.
13. Vanham et al. (2020).
14. https://www.un.org/waterforlifedecade/scarcity.shtml.
15. Ali et al. (2023).

Chapter 13: Trees and their Invisible 'Friends'

1. Edwards et al. (2014).
2. Jacoby et al. (2017).
3. Gunina and Kuzyakov (2022).
4. Callender et al. (2016).
5. Ghedira et al. (2018).
6. Mallet and Roy (2014).
7. http://web.stanford.edu/group/virus/delta/2000/Production.htm.
8. Vlot et al. (2021).
9. Chen et al. (2018).
10. Lindow and Brandt (2003).
11. Wilson (1975).
12. https://www.future-oak.com/.
13. Kennedy and Southward (1984).
14. Denman et al. (2014).
15. Robinson (2023).
16. Knoll (2012).
17. Giovannetti et al. (2006).
18. Sapes et al. (2021).
19. Castro-Delgado et al. (2020).
20. Simard (2021).
21. Guarnizo et al. (2023).
22. Duponnois and Plnechette (2003).
23. Karst et al. (2023).
24. Kiers (2022).
25. Van't Padji et al. (2021a).
26. Hardin (1968).
27. Werner et al. (2015).
28. Kiers et al. (2003).

29. Fellbaum et al. (2014).
30. Foster et al. (2006).
31. Williams (2018).
32. Van't Padji et al. (2021b).
33. Babikova et al. (2013).
34. https://www.spun.earth/.

Chapter 14: Senses and Memory

1. Kurpisz and Pawlowski (2022).
2. Burnett (2011).
3. Lopez et al. (2014) and Khait et al. (2019).
4. Loreto and D'Auria (2022).
5. Amaral et al. (2020).
6. https://membra.info/.
7. Uexküll (1909).
8. Robinson and Cameron (2020).
9. Schnitzler et al. (2003).
10. Fenton (2004).
11. Jones et al. (2016).
12. Liu et al. (2021).
13. Tarnovsky et al. (2023).
14. https://www.cdc.gov/niosh/topics/aircrew/noise.html#:~:text=The%20outside%20of%20aircraft%20engines,insulation%20also%20varies%20among%20aircraft.
15. Wiltschko and Wiltschko (2019).
16. See note 7.
17. Law (1960).
18. Matsuzaki et al. (2006).
19. Rasheed et al. (2023).
20. Karban et al. (2014).
21. Paik and Huq (2019).
22. Fonti et al. (2010).
23. Braeuning et al. (2016).
24. Morley and Lewis (2014).
25. Maclean (1993).
26. Hultsch and Pentz (2014).
27. Dossani et al. (2015).
28. Loftus (2005).
29. Wilson and Golonka (2013).
30. Netea et al. (2019).
31. See note 8.
32. Shapiro and Stoltz (2019).
33. Ciaunica et al. (2023).
34. See note 6.
35. See note 6.
36. https://www.walkingforest.co.uk/.
37. Chen et al. (2022).

38. Gagliano (2018).
39. Gagliano et al. (2014).
40. Gagliano et al. (2016).
41. Khait et al. (2023).

Chapter 15: *Dark Emu*: An Indigenous Perspective

1. https://www.sciencedirect.com/topics/social-sciences/positivism.
2. Pascoe and Gammage (2021).
3. Pascoe (2014).
4. See note 2.
5. Robinson et al. (2021b).
6. https://languageconservancy.org/.
7. Gorenflo et al. (2012).
8. See note 6.
9. Nabhan et al. (2020).
10. https://www.restorationecology.com.au/.

Chapter 16: Restoring Forests in a Changing Climate

1. Feeley et al. (2020).
2. Hartmann et al. (2022).
3. Ward and Aukema (2019).
4. https://thecottonwoodpost.net/2022/03/14/northward-expansion-of-northern-cardinal-carolina-wren-tufted-titmouse-and-red-bellied-woodpecker/.
5. Rosenblad et al. (2023).
6. https://nature.berkeley.edu/news/2023/04/western-forests-are-struggling-keep-climate-change.
7. Goodell, J. (2023).
8. Wolfe (2020).
9. Carrrero et al. (2023).
10. https://yorkshirebylines.co.uk/news/environment/british-trees-under-threat/.
11. https://www.ft.com/content/8bc0b940-7539-4c5a-ad9b-24798808ac2b.
12. https://bsbi.org/wp-content/uploads/dlm_uploads/2023/03/BSBI-Plant-Atlas-2020-press-release-Britain-FINAL.pdf.
13. Williams and Dumroese (2013).
14. Waller et al. (2020).
15. Rohatyn et al. (2022).
16. https://www.pnas.org/post/journal-club/massive-tree-planting-efforts-drylands-can-actually-cause-more-warming.

Chapter 17: Future Forests and Tomorrow's Guardians

1. Palik and D'Amato (2017).
2. MacLean et al. (2022).

3. Heilman et al. (2022).
4. https://www.iaea.org/newscenter/news/what-is-nuclear-fusion
5. https://www.realclearenergy.org/articles/2021/05/12/fusion_ten_times_more_expensive_than_nuclear_power_776839.html
6. Klausen et al. (2022).
7. https://www.dobberendbos.nl/en/
8. https://www.stefanoboeriarchitetti.net/en/project/vertical-forest/
9. Al-Kodmany (2023).
10. Belcher et al. (2018).
11. www.stefanoboeriarchitetti.net/en/project/trudo-vertical-forest/
12. https://www.wri.org/research/global-land-squeeze-managing-growing-competition-land
13. https://findingnature.org.uk/2023/07/24/trees-are-our-closest-neighbours/
14. Birch et al. (2020).
15. Leung et al. (2022).
16. https://www.derby.ac.uk/blog/5-ways-closer-nature/
17. Laaksohariu and Rappe (2017).
18. Lohr et al. (2005).
19. Ng and Leung (2022).

Conclusion: Let Trees Be Thy Teachers

1. Pandey (2008).
2. https://www.theatlantic.com/science/archive/2020/08/chernobyl-fires/615067/

Afterthoughts: Insights from Environmental Psychology

1. Leopold (1949).
2. Barr (2006).
3. Pronin et al. (2002).
4. Hickman et al. (2021).
5. Coffey et al. (2021).
6. Gerber et al. (2023).
7. Whillans and West (2022).
8. Landry et al. (2018).
9. Tian and Liu (2022).
10. Gao et al. (2017).

Bibliography

Ager, A.A., Lasko, R., Myroniuk, V., Zibtsev, S., Day, M.A., Usenia, U., Bogomolov, V., Kovalets, I. and Evers, C.R. 2019. The wildfire problem in areas contaminated by the Chernobyl disaster. *Science of the Total Environment*, 696, 133954.

Agnolín, F.L. and Rozadilla, S. 2018. Phylogenetic reassessment of Pisanosaurus mertii Casamiquela, 1967, a basal dinosauriform from the Late Triassic of Argentina. *Journal of Systematic Palaeontology*, 16(10), pp. 853–879.

Ahirwar, M. and Behera, B.K. 2022. Development of hemp-blended cotton fabrics and analysis on handle behavior, low-stress mechanical and aesthetic properties. *The Journal of the Textile Institute*, 113(5), pp. 934–942.

Al-Kodmany, K. 2023. Greenery-covered tall buildings: a review. *Buildings*, 13(9), p. 2362.

Albright, T.D., Kandel, E.R. and Posner, M.I. 2000. Cognitive neuroscience. *Current Opinion in Neurobiology*, 10(5), pp. 612–624.

Ali, W., Ali, H., Gillani, S., Zinck, P. and Souissi, S. 2023. Polylactic acid synthesis, biodegradability, conversion to microplastics and toxicity: a review. *Environmental Chemistry Letters*, pp. 1–26.

Amaral, J., Ribeyre, Z., Vigneaud, J., Sow, M.D., Fichot, R., Messier, C., Pinto, G., Nolet, P. and Maury, S. 2020. Advances and promises of epigenetics for forest trees. *Forests*, 11(9), p. 976.

Anderson, R.F., Sachs, J.P., Fleisher, M.Q., Allen, K.A., Yu, J., Koutavas, A. and Jaccard, S.L. 2019. Deep-sea oxygen depletion and ocean carbon sequestration during the last ice age. *Global Biogeochemical Cycles*, 33(3), pp. 301–317.

Antonelli, A., Smith, R.J., Fry, C., Simmonds, M.S., Kersey, P.J., Pritchard, H.W., Abbo, M.S., Acedo, C., Adams, J., Ainsworth, A.M. and Allkin, B. 2020. State of the World's Plants and Fungi (Doctoral dissertation, Royal Botanic Gardens (Kew); Sfumato Foundation).

Anthony, M.A., Bender, S.F. and van der Heijden, M.G. 2023. Enumerating soil biodiversity. *Proceedings of the National Academy of Sciences*, 120(33), e2304663120.

Aragão, L.E. 2012. The rainforest's water pump. *Nature*, 489(7415), pp. 217–218.

Armaleo, D., Müller, O., Lutzoni, F., Andrésson, Ó.S., Blanc, G., Bode, H.B., Collart, F.R., Dal Grande, F., Dietrich, F., Grigoriev, I.V. and Joneson, S. 2019. The lichen symbiosis re-viewed through the genomes of *Cladonia grayi* and its algal partner *Asterochloris glomerata*. *BMC genomics*, 20(1), pp. 1–33.

Armenteras, D., Dávalos, L.M., Barreto, J.S., Miranda, A., Hernández-Moreno, A., Zamorano-Elgueta, C., González-Delgado, T.M., Meza-Elizalde, M.C. and Retana, J. 2021. Fire-induced loss of the world's most biodiverse forests in Latin America. *Science Advances*, 7(33), eabd3357.

Atangana, A., Khasa, D., Chang, S., Degrande, A., Atangana, A., Khasa, D., Chang, S. and Degrande, A. 2014. Agroforestry for soil conservation. *Tropical Agroforestry*, pp. 203–216.

Avalos, G., Chacón-Madrigal, E. and Artavia-Rodríguez, L.G. 2021. Invasive plants of Costa Rica current status and research opportunities. *Invasive Alien Species: Observations and Issues from Around the World*, 4, pp. 57–76.

Babikova, Z., Gilbert, L., Bruce, T.J., Birkett, M., Caulfield, J.C., Woodcock, C., Pickett, J.A. and Johnson, D. 2013. Underground signals carried through common mycelial networks warn neighbouring plants of aphid attack. *Ecology Letters*, 16(7), pp. 835–843.

Baker, R.B. 1925. The Men of the Trees. *Quiver*, pp. 223–228.

Barlow, M. 2016. *Boiling Point: Government Neglect, Corporate Abuse, and Canada's Water Crisis*. ECW Press.

Barr, S. 2006. Environmental action in the home: investigating the 'value–action'gap. *Geography*, 91(1), pp. 43–54.

Bastin, J.F., Finegold, Y., Garcia, C., Mollicone, D., Rezende, M., Routh, D., Zohner, C.M. and Crowther, T.W. 2019. The global tree restoration potential. *Science*, 365(6448), pp. 76–79.

Bateman, I. and Balmford, A. 2023. Current conservation policies risk accelerating biodiversity loss. *Nature*, 618(7966), pp. 671–674.

Beans, C. 2023. Can nature inspire sustainable fashion? *Proceedings of the National Academy of Sciences*, 120(20), e2306481120.

Belcher, R.N., Fornasari, L., Menz, S. and Schroepfer, T. 2018. Birds use of vegetated and non-vegetated high-density buildings—a case study of Milan. *Journal of Urban Ecology*, 4(1), juy001.

Bell, T.L., Pattison, A., Keitel, C., Khoddami, A. and Cross, R. 2022. Native grasses as a traditional and emerging source of food. *Australasian Plant Conservation: Journal of the Australian Network for Plant Conservation*, 31(2), pp. 3–6.

Benner, M. 2020. A pedagogical model of CO2 and O2 atmospheric abundances and tree population due to human population. *Journal of Undergraduate Reports in Physics*, 30(1), 100002.

Beresford-Jones, D.G., Arce, S., Whaley, O.Q. and Chepstow-Lusty, A.J. 2009. The role of Prosopis in ecological and landscape change in the Samaca Basin, lower Ica Valley, south coast Peru from the Early Horizon to the Late Intermediate Period. *Latin American Antiquity*, 20(2), pp. 303–332.

Beresford, N.A., Gashchak, S., Wood, M.D. and Barnett, C.L. 2023. Mammals in the Chornobyl Exclusion Zone's Red Forest: a motion-activated camera trap study. *Earth System Science Data*, 15(2), pp. 911–920.

Berry, S., Levinsohn, J. and Pakes, A. 1999. Voluntary export restraints on automobiles: evaluating a trade policy. *American Economic Review*, 89(3), pp. 400–431.

Berry, W. (2018). *The Art of the Commonplace: The agrarian essays of Wendell Berry*. New York: Catapult.

Birch, J., Rishbeth, C. and Payne, S.R. 2020. Nature doesn't judge you–how urban nature supports young people's mental health and wellbeing in a diverse UK city. *Health & Place*, 62, 102296.

Bjørgaas, H.H. 2023. *Secret Life of the City: How Nature Thrives in the Urban Wild*. Greystone Books Ltd.

Bjurhager, I., Gamstedt, E.K., Keunecke, D., Niemz, P. and Berglund, L.A. 2013. Mechanical performance of yew (*Taxus baccata* L.) from a longbow perspective. *Holzforschung*, 67(7), pp. 763–770.

Blair, D.P., Blanchard, W., Banks, S.C. and Lindenmayer, D.B. 2017. Non-linear growth in tree ferns, *Dicksonia antarctica* and *Cyathea australis*. *PLoS One*, 12(5), e0176908.

Bowes, B.G. 2010. *Trees & Forests, a Colour Guide: biology, pathology, propagation, silviculture, surgery, biomes, ecology, and conservation*. CRC Press.

Bowles, A.M., Bechtold, U. and Paps, J. 2020. The origin of land plants is rooted in two bursts of genomic novelty. *Current Biology*, 30(3), pp. 530–536.

Bradby, K., Keesing, A. and Wardell-Johnson, G. 2016. Gondwana Link: connecting people, landscapes, and livelihoods across southwestern Australia. *Restoration Ecology*, 24(6), pp. 827–835.

Bradshaw, C.J., Sodhi, N.S., PEH, K.S.H. and Brook, B.W. 2007. Global evidence that deforestation amplifies flood risk and severity in the developing world. *Global Change Biology*, 13(11), pp. 2379–2395.

Braeuning, A., De Ridder, M., Zafirov, N., García-González, I., Dimitrov, D.P. and Gärtner, H. 2016. Tree-ring features: indicators of extreme event impacts. *Iawa Journal*, 37(2), pp. 206–231.

Bralower, T.J. 2008. Volcanic cause of catastrophe. *Nature*, 454(7202), pp. 285–287.

Breit, S., Kupferberg, A., Rogler, G. and Hasler, G. 2018. Vagus nerve as modulator of the brain–gut axis in psychiatric and inflammatory disorders. *Frontiers in Psychiatry*, p. 44.

Buhrich, A. and Murison, J. 2020. The Western Yalanji dendroglyph: the life and death of an Aboriginal carved tree. *Journal of Community Archaeology & Heritage*, 7(4), pp. 255–271.

Burnett, S. 2011. Perceptual Worlds and Sensory Ecology. *Nature Education Knowledge*, 3(10), 75.

Call, R., Gibson, R. and Kilby, M. 2006. *Pecan production guidelines for small orchards and home yards*.

Callender, K.L., Roy, S., Khasa, D.P., Whyte, L.G. and Greer, C.W. 2016. Actinorhizal alder phytostabilization alters microbial community dynamics in gold mine waste rock from Northern Quebec: a greenhouse study. *PLoS One*, 11(2), e0150181.

Cárdenas, M.L., Pudifoot, B., Narraway, C.L., Pilat, C., Beumer, V. and Hayhow, D.B., Nature-based Solutions Building Urban Resilience for People and the Environment: Tiny forest as a case study. *Quarterly Journal of Forestry*, 116(3), pp. 173–183

Carrero, C., Beckman Bruns, E., Frances, A., Jerome, D., Knapp, W., Meyer, A., Mims, R., Pivorunas, D., Speed, D., Treher Eberly, A. and Westwood, M. 2023. Data sharing for conservation: A standardized checklist of US native tree species and threat assessments to prioritize and coordinate action. *Plants, People, Planet*, 5(4), pp. 600–616.

Carse, A.D. 2006. Trees and trade-offs: perceptions of eucalyptus and native trees in Ecuadorian highland communities. In *Development with Identity: community, culture and sustainability in the Andes* (pp. 103–122). Wallingford UK: CABI Publishing.

Castro-Delgado, A.L., Elizondo-Mesén, S., Valladares-Cruz, Y. and Rivera-Méndez, W. 2020. Wood Wide Web: Communication through the mycorrhizal network. *Revista Tecnología en Marcha*, 33(4), pp. 114–125.

Chau, C., Paulillo, A., Lu, N., Miodownik, M. and Lettieri, P. 2021. The environmental performance of protecting seedlings with plastic tree shelters for afforestation in temperate oceanic regions: A UK case study. *Science of The Total Environment*, 791, p. 148239.

Chavan, S.B., Keerthika, A., Dhyani, S.K., Handa, A.K., Newaj, R. and Rajarajan, K. 2015. National Agroforestry Policy in India: a low hanging fruit. *Current Science*, pp. 1826–1834.

Chen, Y., Wang, J., Yang, N., Wen, Z., Sun, X., Chai, Y. and Ma, Z. 2018. Wheat microbiome bacteria can reduce virulence of a plant pathogenic fungus by altering histone acetylation. *Nature Communications*, 9(1), p. 3429.

Chen, C., Wang, M., Zhu, J., Tang, Y., Zhang, H., Zhao, Q., Jing, M., Chen, Y., Xu, X., Jiang, J. and Shen, Z. 2022. Long-term effect of epigenetic modification in plant–microbe interactions: modification of DNA methylation induced by plant growth-promoting bacteria mediates promotion process. *Microbiome*, 10(1), p. 36.

Cherry, M.A. and Sneirson, J.F. 2012. Chevron, greenwashing, and the myth of 'green oil companies'. *Journal of Energy, Climate, and the Environment*, 3.

Chu, C., Mortimer, P.E., Wang, H., Wang, Y., Liu, X. and Yu, S. 2014. Allelopathic effects of Eucalyptus on native and introduced tree species. *Forest Ecology and Management*, 323, pp. 79–84.

Chua, K.B. 2003. Nipah virus outbreak in Malaysia. *Journal of Clinical Virology*, 26(3), pp. 265–275.

Ciaunica, A., Shmeleva, E.V. and Levin, M. 2023. The brain is not mental! coupling neuronal and immune cellular processing in human organisms. *Frontiers in Integrative Neuroscience*, 17, p. 26.

Clifford, J. and Castonguay, S. 2022. British ghost acres and environmental changes in the Laurentian forest during the nineteenth century. *Journal of Historical Geography*, 78, pp. 126–138.

Climate Focus. (2019). Report on Deforestation. Online: https://climatefocus.com/publications/

Coffey, Y., Bhullar, N., Durkin, J., Islam, M.S. and Usher, K. 2021. Understanding eco-anxiety: A systematic scoping review of current literature and identified knowledge gaps. *The Journal of Climate Change and Health*, 3, p. 100047.

Cohen, J.E. 2003. Human population: the next half century. *Science*, 302(5648), pp. 1172–1175.

Cook-Patton, S.C., Leavitt, S.M., Gibbs, D., Harris, N.L., Lister, K., Anderson-Teixeira, K.J., Briggs, R.D., Chazdon, R.L., Crowther, T.W., Ellis, P.W. and Griscom, H.P. 2020. Mapping carbon accumulation potential from global natural forest regrowth. *Nature*, 585(7826), pp. 545–550.

Cowan, N. 2015. George Miller's magical number of immediate memory in retrospect: observations on the faltering progression of science. *Psychological Review*, 122(3), p. 536.

Crampton, A.B., Stutter, O., Kirby, K.J. and Welch, R.C. 1998. Changes in the composition of Monks Wood National Nature Reserve (Cambridgeshire, UK) 1964–1996. *Arboricultural Journal*, 22(3), pp. 229–245.

Crouzeilles, R., Ferreira, M.S., Chazdon, R.L., Lindenmayer, D.B., Sansevero, J.B., Monteiro, L., Iribarrem, A., Latawiec, A.E. and Strassburg, B.B. 2017. Ecological restoration success is higher for natural regeneration than for active restoration in tropical forests. *Science Advances*, 3(11), e1701345.

Crowther, T.W., Glick, H.B., Covey, K.R., Bettigole, C., Maynard, D.S., Thomas, S.M., Smith, J.R., Hintler, G., Duguid, M.C., Amatulli, G. and Tuanmu, M.N. 2015. Mapping tree density at a global scale. *Nature*, 525(7568), pp. 201–205.

Dai, S., Ma, Y. and Zhang, K. 2022. Land Degradation Caused by Construction Activity: Investigation, Cause and Control Measures. *International Journal of Environmental Research and Public Health*, 19(23), p. 16046.

Das, S. and Vincent, J.R. 2009. Mangroves protected villages and reduced death toll during Indian super cyclone. *Proceedings of the National Academy of Sciences*, 106(18), pp. 7357–7360.

Davis, M.B. 1983. Quaternary history of deciduous forests of eastern North America and Europe. *Annals of the Missouri Botanical Garden*, pp. 550–563.
Dawkins, R. 1976. *The Selfish Gene*. New York: Oxford University Press.
Dawkins, R. 2016. *The Extended Phenotype: The long reach of the gene*. Oxford University Press.
De Min, A., Velicogna, M., Ziberna, L., Chiaradia, M., Alberti, A. and Marzoli, A. 2020. Triassic magmatism in the European Southern Alps as an early phase of Pangea break-up. *Geological Magazine*, 157(11), pp. 1800–1822.
De Nascimento, L., Nogué, S., Naranjo-Cigala, A., Criado, C., McGlone, M., Fernández-Palacios, E. and Fernández-Palacios, J.M. 2020. Human impact and ecological changes during prehistoric settlement on the Canary Islands. *Quaternary Science Reviews*, 239, p. 106332.
de Souza Machado, A.A., Lau, C.W., Kloas, W., Bergmann, J., Bachelier, J.B., Faltin, E., Becker, R., Görlich, A.S. and Rillig, M.C. 2019. Microplastics can change soil properties and affect plant performance. *Environmental Science & Technology*, 53(10), pp. 6044–6052.
Del Rio, D.D.F., Sovacool, B.K., Griffiths, S., Bazilian, M., Kim, J., Foley, A.M. and Rooney, D. 2022. Decarbonizing the pulp and paper industry: A critical and systematic review of sociotechnical developments and policy options. *Renewable and Sustainable Energy Reviews*, 167, p. 112706.
Delate, K., Heller, B. and Shade, J. 2021. Organic cotton production may alleviate the environmental impacts of intensive conventional cotton production. *Renewable Agriculture and Food Systems*, 36(4), pp. 405–412.
Dempsey, N. 2023. The Sheffield street tree dispute: a case of 'business as usual' urban management? *Journal of Environmental Planning and Management*, pp. 1–19.
Denman, S., Brown, N., Kirk, S., Jeger, M. and Webber, J. 2014. A description of the symptoms of Acute Oak Decline in Britain and a comparative review on causes of similar disorders on oak in Europe. *Forestry: An International Journal of Forest Research*, 87(4), pp. 535–551.
Devi, L.S., Kalita, S., Mukherjee, A. and Kumar, S. 2022. Carnauba wax-based composite films and coatings: Recent advancement in prolonging postharvest shelf-life of fruits and vegetables. *Trends in Food Science & Technology*, 129, pp. 296–305.
DiMichele, W.A. 2001. Carboniferous Coal-Swamp Forests. *Palaeobiology II*, pp. 79–82.
Donovan, M. and Monaghan, R. 2021. Impacts of grazing on ground cover, soil physical properties and soil loss via surface erosion: a novel geospatial modelling approach. *Journal of environmental Management*, 287, p. 112206.
Dossani, R.H., Missios, S. and Nanda, A. 2015. The legacy of Henry Molaison (1926–2008) and the impact of his bilateral mesial temporal lobe surgery on the study of human memory. *World Neurosurgery*, 84(4), pp. 1127–1135.
Duarte, C.M., Gattuso, J.P., Hancke, K., Gundersen, H., Filbee-Dexter, K., Pedersen, M.F., Middelburg, J.J., Burrows, M.T., Krumhansl, K.A., Wernberg, T. and Moore, P. 2022. Global estimates of the extent and production of macroalgal forests. *Global Ecology and Biogeography*, 31(7), pp. 1422–1439.
Duponnois, R. and Plenchette, C. 2003. A mycorrhiza helper bacterium enhances ectomycorrhizal and endomycorrhizal symbiosis of Australian Acacia species. *Mycorrhiza*, 13(2), pp. 85–91.
Edwards, A., Irvine-Fynn, T., Mitchell, A.C. and Rassner, S.M. 2014. A germ theory for glacial systems? *Wiley Interdisciplinary Reviews: Water*, 1(4), pp. 331–340.

Elbert, W., Weber, B., Burrows, S., Steinkamp, J., Büdel, B., Andreae, M.O. and Pöschl, U. 2012. Contribution of cryptogamic covers to the global cycles of carbon and nitrogen. *Nature Geoscience*, 5(7), pp. 459–462.

Evangeliou, N., Balkanski, Y., Cozic, A., Hao, W.M. and Møller, A.P. 2014. Wildfires in Chernobyl-contaminated forests and risks to the population and the environment: A new nuclear disaster about to happen? *Environment international*, 73, pp. 346–358.

Evans, S. 2010. *The Green Republic: A conservation history of Costa Rica*. University of Texas Press.

Feeley, K.J., Bravo-Avila, C., Fadrique, B., Perez, T.M. and Zuleta, D. 2020. Climate-driven changes in the composition of New World plant communities. *Nature Climate Change*, 10(10), pp. 965–970.

Fellbaum, C.R., Mensah, J.A., Cloos, A.J., Strahan, G.E., Pfeffer, P.E., Kiers, E.T. and Bücking, H. 2014. Fungal nutrient allocation in common mycorrhizal networks is regulated by the carbon source strength of individual host plants. *New Phytologist*, 203(2), pp. 646–656.

Fenton, M.B. 2004, Bat natural history and echolocation. *Bat echolocation research: tools, techniques and analysis*. 2.

Ferrante, L. and Fearnside, P.M. 2019. Brazil's new president and 'ruralists' threaten Amazonia's environment, traditional peoples and the global climate. *Environmental Conservation*, 46(4), pp. 261–263.

Field, C.B., Behrenfeld, M.J., Randerson, J.T. and Falkowski, P. 1998. Primary production of the biosphere: integrating terrestrial and oceanic components. *Science*, 281(5374), pp. 237–240.

Filbee-Dexter, K. and Wernberg, T. 2018. Rise of turfs: a new battlefront for globally declining kelp forests. *BioScience*, 68(2), pp. 64–76.

Fitz-James, M.H. and Cavalli, G. 2022. Molecular mechanisms of transgenerational epigenetic inheritance. *Nature Reviews Genetics*, 23(6), pp. 325–341.

Fleischman, F., Basant, S., Chhatre, A., Coleman, E.A., Fischer, H.W., Gupta, D., Güneralp, B., Kashwan, P., Khatri, D., Muscarella, R. and Powers, J.S. 2020. Pitfalls of tree planting show why we need people-centered natural climate solutions. *BioScience*, 70(11), pp. 947–950.

Fletcher, M.S., Hamilton, R., Dressler, W. and Palmer, L. 2021. Indigenous knowledge and the shackles of wilderness. *Proceedings of the National Academy of Sciences*, 118(40), e2022218118.

Fonti, P., von Arx, G., García-González, I., Eilmann, B., Sass-Klaassen, U., Gärtner, H. and Eckstein, D. 2010. Studying global change through investigation of the plastic responses of xylem anatomy in tree rings. *New Phytologist*, 185(1), pp. 42–53.

Forzieri, G., Girardello, M., Ceccherini, G., Spinoni, J., Feyen, L., Hartmann, H., Beck, P.S., Camps-Valls, G., Chirici, G., Mauri, A. and Cescatti, A. 2021. Emergent vulnerability to climate-driven disturbances in European forests. *Nature Communications*, 12(1), p. 1081.

Foster, K.R., Wenseleers, T. and Ratnieks, F.L. 2006. Kin selection is the key to altruism. *Trends in Ecology & Evolution*, 21(2), pp. 57–60.

Fremout, T., Thomas, E., Gaisberger, H., Van Meerbeek, K., Muenchow, J., Briers, S., Gutierrez-Miranda, C.E., Marcelo-Peña, J.L., Kindt, R., Atkinson, R. and Cabrera, O. 2020. Mapping tree species vulnerability to multiple threats as a guide to restoration and conservation of tropical dry forests. *Global Change Biology*, 26(6), pp. 3552–3568.

Friis, E.M., Crane, P.R. and Pedersen, K.R. 2019. Extinct diversity among Early Cretaceous angiosperms: mesofossil evidence of early Magnoliales from Portugal. *International Journal of Plant Sciences*, 180(2), pp. 93–127.

Fuller, D.Q. 2011. Pathways to Asian civilizations: tracing the origins and spread of rice and rice cultures. *Rice*, 4(3), pp. 78–92.

Gagliano, M. 2013. Green symphonies: a call for studies on acoustic communication in plants. *Behavioral Ecology*, 24(4), pp. 789–796.

Gagliano, M., Renton, M., Depczynski, M. and Mancuso, S. 2014. Experience teaches plants to learn faster and forget slower in environments where it matters. *Oecologia*, 175, pp. 63–72.

Gagliano, M. 2018. *Thus Spoke the Plant: A remarkable journey of groundbreaking scientific discoveries and personal encounters with plants*. North Atlantic Books.

Gammage, B. 2013. *The Biggest Estate on Earth: How Aborigines Made Australia*. Allen & Unwin 2011.

Gao, J., Huang, Z. and Zhang, C. 2017. Tourists' perceptions of responsibility: an application of norm-activation theory. *Journal of Sustainable Tourism*, 25(2), pp. 276–291.

Gerber, L.R., Reeves-Blurton, Z., Gueci, N., Iacona, G.D., Beaudette, J.A. and Pipe, T. 2023. Practicing mindfulness in addressing the biodiversity crisis. *Conservation Science and Practice*, e12945.

Ghedira, K., Harigua-Souiai, E., Ben Hamda, C., Fournier, P., Pujic, P., Guesmi, S., Guizani, I., Miotello, G., Armengaud, J., Normand, P. and Sghaier, H. 2018. The PEG-responding desiccome of the alder microsymbiont *Frankia alni*. *Scientific Reports*, 8(1), p. 759.

Giovannetti, M., Avio, L., Fortuna, P., Pellegrino, E., Sbrana, C. and Strani, P. 2006. At the root of the wood wide web: self recognition and nonself incompatibility in mycorrhizal networks. *Plant Signaling & Behavior*, 1(1), pp. 1–5.

Goodell, J. (2023). *The Heat Will Kill You First*. Little, Brown and Company.

Gorenflo, L.J., Romaine, S., Mittermeier, R.A. and Walker-Painemilla, K. 2012. Co-occurrence of linguistic and biological diversity in biodiversity hotspots and high biodiversity wilderness areas. *Proceedings of the National Academy of Sciences*, 109(21), pp. 8032–8037.

Graziosi, I., Tembo, M., Kuate, J. and Muchugi, A. 2020. Pests and diseases of trees in Africa: a growing continental emergency. *Plants, People, Planet*, 2(1), pp. 14–28.

Griffiths, J. 2007. *Wild: An elemental journey*. London: Penguin UK.

Guarnizo, Á.L., Navarro-Ródenas, A., Calvo-Polanco, M., Marqués-Gálvez, J.E. and Morte, A. 2023. A mycorrhizal helper bacterium alleviates drought stress in mycorrhizal *Helianthemum almeriense* plants by regulating water relations and plant hormones. *Environmental and Experimental Botany*, 207, p. 105228.

Gunina, A. and Kuzyakov, Y. 2022. From energy to (soil organic) matter. *Global Change Biology*, 28(7), pp. 2169–2182.

Hallam, A. 1982. *The Jurassic Climate. Studies in Geophysics: Climate in Earth History*. National Academy Press. Washington, DC, pp. 159–163.

Hanley, P. 1983. Richard St Barbe Baker: Man of the Trees. *The Structurist*, (23), p. 22.

Harada, A.Y., Adis, J., Stork, N.E., Adis, J. and Didham, R.K. 1997. *Canopy Arthropods*. London: Chapman & Hall, 1997.

Hardin, G. 1968. The tragedy of the commons: the population problem has no technical solution; it requires a fundamental extension in morality. *Science*, 162(3859), pp. 1243–1248.

Hartmann, H., Bastos, A., Das, A.J., Esquivel-Muelbert, A., Hammond, W.M., Martínez-Vilalta, J., McDowell, N.G., Powers, J.S., Pugh, T.A., Ruthrof, K.X. and Allen, C.D. 2022. Climate change risks to global forest health: emergence of unexpected events of elevated tree mortality worldwide. *Annual Review of Plant Biology*, 73, pp. 673–702.

He, K., Lu, H., Sun, G., Wang, Y., Zheng, Y., Zheng, H., Lei, S., Li, Y. and Zhang, J. 2022. Dynamic interaction between deforestation and rice cultivation during the Holocene in the lower Yangtze River, China. *Frontiers in Earth Science*, 10, p. 849501.

Hebelstrup, K.H., Azariadis, A., Cordes, A., Henriksen, P.S. and Brinch-Pedersen, H. 2023. Prehistoric plant exploitation and domestication: an inspiration for the science of de novo domestication in present times. *Plants*, 12(12), p. 2310.

Heilman, K.A., Dietze, M.C., Arizpe, A.A., Aragon, J., Gray, A., Shaw, J.D., Finley, A.O., Klesse, S., DeRose, R.J. and Evans, M.E. 2022. Ecological forecasting of tree growth: Regional fusion of tree-ring and forest inventory data to quantify drivers and characterize uncertainty. *Global Change Biology*, 28(7), pp. 2442–2460.

Herron, S. 1998. The Economic botany of *Manilkara zapota* (L.) Van Royen. *Ethnobotanical Leaflets*, 1999(2), p. 2.

Hickman, C., Marks, E., Pihkala, P., Clayton, S., Lewandowski, R.E., Mayall, E.E., Wray, B., Mellor, C. and van Susteren, L. 2021. Climate anxiety in children and young people and their beliefs about government responses to climate change: a global survey. *The Lancet Planetary Health*, 5(12), e863-e873.

Hoang, N.T. and Kanemoto, K. 2021. Mapping the deforestation footprint of nations reveals growing threat to tropical forests. *Nature Ecology & Evolution*, 5(6), pp. 845–853.

Hultsch, D.F. and Pentz, C.A. 2014. Encoding, storage, and retrieval in adult memory: The role of model assumptions. *In New Directions in Memory and Aging (PLE: Memory)* (pp. 73–94). Psychology Press.

Hussain, M., u Rahman, M.K., Mishra, R.C. and Van Der Straeten, D. 2023. Plants can talk: a new era in plant acoustics. *Trends in Plant Science*, 28(9), pp. 987–990.

Irwin, A., Geschke, A., Brooks, T.M., Siikamaki, J., Mair, L. and Strassburg, B.B. 2022. Quantifying and categorising national extinction-risk footprints. *Scientific Reports*, 12(1), p. 5861.

Jacoby, R., Peukert, M., Succurro, A., Koprivova, A. and Kopriva, S. 2017. The role of soil microorganisms in plant mineral nutrition – current knowledge and future directions. *Frontiers in Plant Science*, 8, p. 1617.

Jiang, X., Chang, Y., Zhang, T., Qiao, Y., Klobučar, G. and Li, M. 2020. Toxicological effects of polystyrene microplastics on earthworm (*Eisenia fetida*). *Environmental Pollution*, 259, p. 113896.

Johnson, L., Evans, J., Montgomery, J. and Chenery, C. 2022. The forest effect: Biosphere 87Sr/86Sr shifts due to changing land use and the implications for migration studies. *Science of the Total Environment*, 839, p. 156083.

Jones, P.L., Page, R.A. and Ratcliffe, J.M. 2016. To scream or to listen? Prey detection and discrimination in animal-eating bats. *Bat Bioacoustics*, pp. 93–116.

Kanchan, J., Ganaraj, K., Hegde, M.N. and Shetty, S.S. 2023. Automatic Seed Dispensing System Using Drone Technology. In *Applying Drone Technologies and Robotics for Agricultural Sustainability* (pp. 95–107). IGI Global.

Kappelle, M. 2016. Costa Rica's ecosystems: setting the stage. *Costa Rican Ecosystems*, p. 3.

Karban, R., Yang, L.H. and Edwards, K.F. 2014. Volatile communication between plants that affects herbivory: A meta-analysis. *Ecology Letters*, 17(1), pp. 44–52.

Karst, J., Jones, M.D. and Hoeksema, J.D. 2023. Positive citation bias and overinterpreted results lead to misinformation on common mycorrhizal networks in forests. *Nature Ecology & Evolution*, 7(4), pp. 501–511.

Kashparov, V., Yoschenko, V., Levchuk, S., Bugai, D., Van Meir, N., Simonucci, C. and Martin-Garin, A. 2012. Radionuclide migration in the experimental polygon of the Red Forest waste site in the Chernobyl zone–Part 1: Characterization of the waste trench, fuel particle transformation processes in soils, biogenic fluxes and effects on biota. *Applied Geochemistry*, 27(7), pp. 1348–1358.

Kennedy, C.E.J. and Southwood, T.R.E. 1984. The number of species of insects associated with British trees: a re-analysis. *The Journal of Animal Ecology*, 53, pp. 455–478.

Khait, I., Obolski, U., Yovel, Y. and Hadany, L. 2019, August. Sound perception in plants. In *Seminars in Cell & Developmental Biology*, 92, pp. 134–138. Academic Press.

Khait, I., Lewin-Epstein, O., Sharon, R., Saban, K., Goldstein, R., Anikster, Y., Zeron, Y., Agassy, C., Nizan, S., Sharabi, G. and Perelman, R. 2023. Sounds emitted by plants under stress are airborne and informative. *Cell*, 186(7), pp. 1328–1336.

Kiers, E.T., Rousseau, R.A., West, S.A. and Denison, R.F. 2003. Host sanctions and the legume–rhizobium mutualism. *Nature*, 425(6953), pp. 78–81.

Kiers, T. 2022. Toby Kiers. *Current Biology*, 32(4), pp.R147-R148.

Kimmerer, R. 2013. *Braiding Sweetgrass: Indigenous Wisdom, Scientific Knowledge and the Teachings of Plants*. Minneapolis: Milkweed Editions.

Kimmerer, R.W. 2021. *Gathering Moss: A Natural and Cultural History of Mosses*. Penguin UK.

Kings, M. 2022. Ash dieback: the fatal fungus costing the Earth Environment| Forests. Online. https://www.the-kingfisher.org/environment/forests/europe/ash_dieback.html. Accessed on 11.02.24.

Klausen, S.H., Emiliussen, J., Christiansen, R., Hasandedic-Dapo, L. and Engelsen, S. 2022. The many faces of hedonic adaptation. *Philosophical Psychology*, 35(2), pp. 253–278.

Klein Goldewijk, K., Beusen, A., Doelman, J. and Stehfest, E. 2016. New anthropogenic land use estimates for the Holocene; *HYDE 3.2, Earth Sysems Science Data*. 9(2), pp. 927–953.

Knoll, A.H. 2012. Lynn Margulis, 1938–2011. *Proceedings of the National Academy of Sciences*, 109(4), pp. 1022–1022.

Knudson, K.J., Williams, S.R., Osborn, R., Forgey, K. and Williams, P.R. 2009. The geographic origins of Nasca trophy heads using strontium, oxygen, and carbon isotope data. *Journal of Anthropological Archaeology*, 28(2), pp. 244–257.

Kopfman, J.E., Smith, S.W., Ah Yun, J.K. and Hodges, A. 1998. Affective and cognitive reactions to narrative versus statistical evidence organ donation messages. *Journal of Applied Communications Research*, 26, pp. 279–300.

Krám, P., Oulehle, F., Štědrá, V., Hruška, J., Shanley, J.B., Minocha, R. and Traister, E. 2009. Geoecology of a forest watershed underlain by serpentine in central Europe. *Northeastern Naturalist*, 16(sp5), pp. 309–328.

Kremer, A. and Hipp, A.L. 2020. Oaks: an evolutionary success story. *New Phytologist*, 226(4), pp. 987–1011.

Kumar, V. 2017. Recycling of waste and used papers: a useful contribution in conservation of environment: a case study. *Asian Journal of Water, Environment and Pollution*, 14(4), pp. 31–36.

Kurek, P., Dobrowolska, D. and Wiatrowska, B. 2019. Dispersal distance and burial mode of acorns in Eurasian Jays *Garrulus glandarius* in European temperate forests. *Acta Ornithologica*, 53(2), pp. 155–162.

Kurpisz, B. and Pawłowski, T.A. 2022. Epigenetic mechanisms of tree responses to climatic changes. *International Journal of Molecular Sciences*, 23(21), p. 13412.

Laaksoharju, T. and Rappe, E. 2017. Trees as affordances for connectedness to place – a framework to facilitate children's relationship with nature. *Urban Forestry & Urban Greening*, 28, pp. 150–159.

Landry, N., Gifford, R., Milfont, T.L., Weeks, A. and Arnocky, S. 2018. Learned helplessness moderates the relationship between environmental concern and behavior. *Journal of Environmental Psychology*, 55, pp. 18–22.

Landsman, A.P., Schmit, J.P. and Matthews, E.R. 2021. Invasive plants differentially impact forest invertebrates, providing taxon-specific benefits by enhancing structural complexity. *Frontiers in Ecology and Evolution*, 9, p. 682140.

Lanthorn, K.R. 2013. It's all about the green: The economically driven greenwashing practices of Coca-Cola. *Augsburg Honors Review*, 6(1), p. 13.

Law. C. 1960. The great gray owl of the woodlands. *Blue Jay*, 18(1), pp. 14–17.

Lee, S., Lee, R.J. and Scherr, S. 2023. How tree canopy cover can reduce urban suicide attempts: a geospatial analysis of the moderating role of area deprivation. *Landscape and Urban Planning*, 230, p. 104606.

Lees, A.C., Haskell, L., Allinson, T., Bezeng, S.B., Burfield, I.J., Renjifo, L.M., Rosenberg, K.V., Viswanathan, A. and Butchart, S.H. 2022. State of the world's birds. *Annual Review of Environment and Resources*, 47, pp. 231–260.

Lehmann, P., Boratyński, Z., Mappes, T., Mousseau, T.A. and Møller, A.P. 2016. Fitness costs of increased cataract frequency and cumulative radiation dose in natural mammalian populations from Chernobyl. *Scientific Reports*, 6(1), p. 19974.

Leigh Jr, E.G. 2010. The group selection controversy. *Journal of Evolutionary Biology*, 23(1), pp. 6–19.

Leopold, A. 1949. *A Sand County Almanac (outdoor essays & reflections)*. Ballantine (reprint 1966 and 1990).

Leung, G.Y., Hazan, H. and Chan, C.S. 2022. Exposure to nature in immersive virtual reality increases connectedness to nature among people with low nature affinity. *Journal of Environmental Psychology*, 83, 101863.

Lewis, H. 2022. *Mini-Forest Revolution: Using the Miyawaki Method to Rapidly Rewild the World*. Chelsea Green Publishing.

Liao, J. and Chen, Q. 2021. Biodegradable plastics in the air and soil environment: low degradation rate and high microplastics formation. *Journal of Hazardous Materials*, 418, 126329.

Li, Q., Kobayashi, M., Wakayama, Y., Inagaki, H., Katsumata, M., Hirata, Y., Hirata, K., Shimizu, T., Kawada, T., Park, B.J. and Ohira, T. 2009. Effect of phytoncide from trees on human natural killer cell function. *International Journal of Immunopathology and Pharmacology*, 22(4), pp. 951–959.

Li, Q. 2018. *Forest Bathing: How trees can help you find health and happiness*. Penguin.

Li, Q. 2022. Effects of forest environment (Shinrin-yoku/Forest bathing) on health promotion and disease prevention – the establishment of 'Forest Medicine'. *Environmental Health and Preventive Medicine*, 27, pp. 43–43.

Lindow, S.E. and Brandl, M.T. 2003. Microbiology of the phyllosphere. *Applied and Environmental Microbiology*, 69(4), pp. 1875–1883.

Lippke, B., Puettmann, M., Oneil, E. and Dearing Oliver, C. 2021. The plant a trillion trees campaign to reduce global warming–fleshing out the concept. *Journal of Sustainable Forestry*, 40(1), pp. 1–31.

Liu, Z., Chen, P., Li, Y.Y., Li, M.W., Liu, Q., Pan, W.L., Xu, D.M., Bai, J., Zhang, L.B., Tang, J. and Shi, P. 2021. Cochlear hair cells of echolocating bats are immune to intense noise. *Journal of Genetics and Genomics*, 48(11), pp. 984–993.

Loftus, E.F. 2005. Planting misinformation in the human mind: A 30-year investigation of the malleability of memory. *Learning & Memory*, 12(4), pp. 361–366.

Lohr, V.I. and Pearson-Mims, C.H. 2005. Children's active and passive interactions with plants influence their attitudes and actions toward trees and gardening as adults. *HortTechnology*, 15(3), pp. 472–476.

Loidi, J. and Fernández-González, F. 2012. Potential natural vegetation: reburying or reboring? *Journal of Vegetation Science*, 23(3), pp. 596–604.

Long, J.W., Lake, F.K. and Goode, R.W. 2021. The importance of Indigenous cultural burning in forested regions of the Pacific West, USA. *Forest Ecology and Management*, 500, p. 119597.

Lopez, D., Tocquard, K., Venisse, J.S., Legué, V. and Roeckel-Drevet, P. 2014. Gravity sensing, a largely misunderstood trigger of plant-orientated growth. *Frontiers in Plant Science*, 5, p. 610.

Loreto, F. and D'Auria, S. 2022. How do plants sense volatiles sent by other plants? *Trends in Plant Science*, 27(1), pp. 29–38.

Lydia, N., Basamba, T.A., Nyakoojo, C., Mustafa, A.S., Saidi, N., Mutumba, G.M. and Ssenku, J.E. 2023. Ecological distribution and genetic diversity of *Azolla* in Uganda. *BMC Plant Biology*, 23(1), p. 131.

Maathai, W. 2003. *The Green Belt Movement: Sharing the approach and the experience*. Lantern Books.

Maclean, D. 1993. Keynote Address ('Using Water Wisely') to IWEM's Annual Symposium on 'The Management of Scarce Water Resources'. *Water and Environment Journal*, 7(2), pp. 196–198.

MacLean, D.A., Taylor, A.R., Neily, P.D., Steenberg, J.W., Basquill, S.P., Quigley, E., Boone, C.K., Oikle, M., Bush, P. and Stewart, B. 2022. Natural disturbance regimes for implementation of ecological forestry: a review and case study from Nova Scotia, Canada. *Environmental Reviews*, 30(1), pp. 128–158.

Malhi, Y. 2018. Ancient deforestation in the green heart of Africa. *Proceedings of the National Academy of Sciences*, 115(13), pp. 3202–3204.

Mallet, P.L. and Roy, S. 2014. The symbiosis between Frankia and alder shrubs results in a tolerance of the environmental stress associated with tailings from the Canadian oil sands industry. *Journal of Petroleum and Environmental Biotechnology*, 5(3), pp. 1–9.

Margulis, L. and Sagan, D. 1995. *What is Life?* Berkeley.

Matsuzaki, J., Masumori, M. and Tange, T. 2006. Stem phototropism of trees: a possible significant factor in determining stem inclination on forest slopes. *Annals of Botany*, 98(3), pp. 573–581.

Mavrogianni, A., Taylor, J., Milojevic, A., Oikonomou, E., Heaviside, C., Tsoulou, I., Petrou, G., Symonds, P., Davies, M. and Wilkinson, P. 2022. Heat-Related Mortality in London. In *Urban Overheating: Heat Mitigation and the Impact on Health* (pp. 171–180). Singapore: Springer Nature Singapore.

McAnany, P.A. and Yoffee, N., eds. 2009. *Questioning Collapse: human resilience, ecological vulnerability, and the aftermath of empire*. Cambridge University Press.

Mda, Z. 2016. Green Citizens and Problematic Populations in *The Green State in Africa*, p. 108. New Haven: Yale University Press.

Meier, R., Schwaab, J., Seneviratne, S.I., Sprenger, M., Lewis, E. and Davin, E.L. 2021. Empirical estimate of forestation-induced precipitation changes in Europe. *Nature Geoscience*, 14(7), pp. 473–478.

Meloni, M. 2022. Molecular Epigenetics, the Biology of Memory, and Biology as Memory. *Cultural Memory: From the Sciences to the Humanities*, p. 3.

Men of the Trees. 1933. *Nature*. 132, 477. https://doi.org/10.1038/132477a0

Meyer-Berthaud, B., Scheckler, S.E. and Wendt, J. 1999. Archaeopteris is the earliest known modern tree. *Nature*, 398(6729), pp. 700–701.

Mitchell, F.J. 2005. How open were European primeval forests? Hypothesis testing using palaeoecological data. *Journal of Ecology*, 93(1), pp. 168–177.

Mitchell, R.J., Bellamy, P.E., Broome, A., Ellis, C.J., Hewison, R.L., Iason, G.R., Littlewood, N.A., Newey, S., Pozsgai, G., Ray, D. and Stockan, J.A. 2022. Cumulative impact assessments of multiple host species loss from plant diseases show disproportionate reductions in associated biodiversity. *Journal of Ecology*, 110(1), pp. 221–231.

Miyawaki, A. 1993. Restoration of native forests from Japan to Malaysia. In *Restoration of Tropical Forest Ecosystems: Proceedings of the Symposium held on October 7–10, 1991* (pp. 5–24). Springer Netherlands.

Møller, A.P., Bonisoli-Alquati, A., Rudolfsen, G. and Mousseau, T.A. 2011. Chernobyl birds have smaller brains. *PLoS One*, 6(2), e16862.

Monbiot, G. 2022. *Regenesis: Feeding the world without devouring the planet*. Penguin.

Morley, N.J. and Lewis, J.W. 2014. Extreme climatic events and host–pathogen interactions: The impact of the 1976 drought in the UK. *Ecological Complexity*, 17, pp. 1–19.

Murre, J.M. and Dros, J. 2015. Replication and analysis of Ebbinghaus' forgetting curve. *PLoS One*, 10(7), e0120644.

Nabhan, G.P., Orlando, L., Smith Monti, L. and Aronson, J. 2020. Hands-on ecological restoration as a nature-based health intervention: Reciprocal restoration for people and ecosystems. *Ecopsychology*, 12(3), pp. 195–202.

Ndangalasi, H.J., Martínez-Garza, C., Harjo, T.C., Pedigo, C.A., Wilson, R.J. and Cordeiro, N.J. 2021. Seedling recruitment under isolated trees in a tea plantation provides a template for forest restoration in eastern Africa. *PLoS One*, 16(5), e0250859.

Netea, M.G., Schlitzer, A., Placek, K., Joosten, L.A. and Schultze, J.L. 2019. Innate and adaptive immune memory: an evolutionary continuum in the host's response to pathogens. *Cell Host & Microbe*, 25(1), pp. 13–26.

Ng, H.K.S. and Leung, A.N.M. 2022. Nature connectedness and nature exposure interactively influence social dominance orientation and policy support for marginalized groups during the COVID-19 pandemic. *Environment and Behavior*, 54(2), pp. 412–449.

Paik, I. and Huq, E. 2019, August. Plant photoreceptors: Multi-functional sensory proteins and their signaling networks. In *Seminars in Cell & Developmental Biology* (Vol. 92, pp. 114–121). Academic Press.

Palik, B.J. and D'Amato, A.W. 2017. Ecological forestry: much more than retention harvesting. *Journal of Forestry*, 115(1), pp. 51–53.

Pandey, M. 2008. Birbal: a pioneer in lateral thinking. *Growth-Journal of the Management Training Institute*, 36(3), pp. 1–6.

Pascoe, B. 2014. *Dark Emu, Black Seeds: Agriculture or Accident?* Magabala Books.

Pascoe, B. and Gammage, B. 2021. *First Knowledges Country: Future Fire, Future Farming*. Thames & Hudson Australia.

Patiño, J. and Vanderpoorten, A. 2018. Bryophyte biogeography. *Critical Reviews in Plant Sciences*, 37(2-3), pp. 175–209.

Pearce, E.A., Mazier, F., Normand, S., Fyfe, R., Andrieu, V., Bakels, C., Balwierz, Z., Bińka, K., Boreham, S., Borisova, O.K. and Brostrom, A. 2023. Substantial light woodland and open vegetation characterized the temperate forest biome before *Homo sapiens*. *Science Advances*, 9(45), eadi9135.

Pearce, F. (2023). Climate change could expand forests. Online: https://www.science.org/content/article/trees-help-curb-climate-change-can-also-contribute-warming-reducing-earths-reflectivity

Pearsall, D.M. 2008. Plant domestication and the shift to agriculture in the Andes. In *The Handbook of South American Archaeology* (pp. 105–120). New York, NY: Springer New York.

Pe'Er, G., Zinngrebe, Y., Moreira, F., Sirami, C., Schindler, S., Müller, R., Bontzorlos, V., Clough, D., Bezák, P., Bonn, A. and Hansjürgens, B. 2019. A greener path for the EU Common Agricultural Policy. *Science*, 365(6452), pp. 449–451.

Pempel, T.J. 1987. The unbundling of "Japan, Inc.": The changing dynamics of Japanese policy formation. *Journal of Japanese Studies*, 13(2), pp. 271–306.

Penfield, W. 1952. Memory mechanisms. *AMA Archives of Neurology & Psychiatry*, 67(2), pp. 178–198.

Pio, C.A., Silva, P.A., Cerqueira, M.A. and Nunes, T.V. 2005. Diurnal and seasonal emissions of volatile organic compounds from cork oak (*Quercus suber*) trees. *Atmospheric Environment*, 39(10), pp. 1817–1827.

Piperno, D.R., McMichael, C.H., Pitman, N.C., Andino, J.E.G., Ríos Paredes, M., Heijink, B.M. and Torres-Montenegro, L.A. 2021. A 5,000-year vegetation and fire history for tierra firme forests in the Medio Putumayo-Algodón watersheds, northeastern Peru. *Proceedings of the National Academy of Sciences*, 118(40), e2022213118.

Poddar, S. 2021. Miyawaki technique of afforestation. *Krishi Science-eMagazine Agricultural Sciences*, 2, pp. 1–3.

Poore, R.Z., Williams, R.S., Jr., and Tracey, Christopher, 2000, Sea level and climate: U.S. Geological Survey Fact Sheet 002–00, at: https://pubs.usgs.gov/fs/fs2-00/.

Prach, K., Tichý, L., Lencová, K., Adámek, M., Koutecký, T., Sádlo, J., Bartošová, A., Novák, J., Kovář, P., Jírová, A. and Šmilauer, P. 2016. Does succession run towards potential natural vegetation? An analysis across seres. *Journal of Vegetation Science*, 27(3), pp. 515–523.

Pronin, E., Lin, D.Y. and Ross, L. 2002. The bias blind spot: Perceptions of bias in self versus others. *Personality and Social Psychology Bulletin*, 28(3), pp. 369–381.

Rackham, O. 1986. *The History of the Countryside*. London: JM Dent.

Rahman, M.A., Stratopoulos, L.M., Moser-Reischl, A., Zölch, T., Häberle, K.H., Rötzer, T., Pretzsch, H. and Pauleit, S. 2020. Traits of trees for cooling urban heat islands: A meta-analysis. *Building and Environment*, 170, p. 106606.

Rahman, M.B., (2022). Greenwashing in Bangladesh: How to Combat its Spread. *International Journal of Social Sciences: Current and Future Research Trends* (IJSSCFRT), 15, No 1, pp. 139–153

Rai, P.K. and Singh, J.S. 2020. Invasive alien plant species: Their impact on environment, ecosystem services and human health. *Ecological Indicators*, 111, p. 106020.

Raines, K.E., Whitehorn, P.R., Copplestone, D. and Tinsley, M.C. 2020. Chernobyl-level radiation exposure damages bumblebee reproduction: a laboratory experiment. *Proceedings of the Royal Society B*, 287(1937), p. 20201638.

Ramage, M.H., Burridge, H., Busse-Wicher, M., Fereday, G., Reynolds, T., Shah, D.U., Wu, G., Yu, L., Fleming, P., Densley-Tingley, D. and Allwood, J. 2017. The wood from the trees: The use of timber in construction. *Renewable and Sustainable Energy Reviews*, 68, pp. 333–359.

Ramírez-Toro, W., Torres-Miranda, A., González-Rodríguez, A., Ruiz-Sanchez, E., Luna-Vega, I. and Oyama, K. 2017. A multicriteria analysis for prioritizing areas for conservation of oaks (Fagaceae: *Quercus*) in Oaxaca, southern Mexico. *Tropical Conservation Science*, 10, p. 1940082917714227.

Rashad, S.M. and Hammad, F.H. 2000. Nuclear power and the environment: comparative assessment of environmental and health impacts of electricity-generating systems. *Applied Energy*, 65(1-4), pp. 211–229.

Rasheed, M.U., Brosset, A., and Blande, J.D. 2023. Tree communication: the effects of 'wired' and 'wireless' channels on interactions with herbivores. *Forest Entomology*. 9, 33–47.

Raven, P. and Wackernagel, M. 2020. Maintaining biodiversity will define our long-term success. *Plant Diversity*, 42(4), pp. 211–220.

Richards, T. 2011. A late nineteenth-century map of an Australian Aboriginal fishery at Lake Condah. *Australian Aboriginal Studies*, (2), pp. 64–87.

Ritchie, H. and Roser, M. 2018. *Plastic Pollution*. Our World in Data.

Ritchie, H. and Roser, M. 2021. *Forests and Deforestation*. Our world in data.

Rival, D., Hass, G. and Tropea, C. 2011. Recovery of energy from leading-and trailing-edge vortices in tandem-airfoil configurations. *Journal of Aircraft*, 48(1), pp. 203–211.

Roberts, E. and Todd, C. 2017. Were termites the world's first farmers? *Australasian Science*, 38(2), pp. 22–24.

Robinson, J.M. and Breed, M.F. 2019. Green prescriptions and their co-benefits: Integrative strategies for public and environmental health. *Challenges*, 10(1), p. 9.

Robinson, J.M., Jorgensen, A., Cameron, R. and Brindley, P. 2020. Let nature be thy medicine: a socioecological exploration of green prescribing in the UK. *International Journal of Environmental Research and Public Health*, 17(10), p. 3460.

Robinson, J.M., Cando-Dumancela, C., Liddicoat, C., Weinstein, P., Cameron, R. and Breed, M.F. 2020. Vertical stratification in urban green space aerobiomes. *Environmental Health Perspectives*, 128(11), p. 117008.

Robinson J.M., Cameron, R. 2020. The holobiont blindspot: relating host-microbiome interactions to cognitive biases and the concept of the *umwelt*. *Frontiers in Psychology*, 11. doi: 10.3389/fpsyg.2020.591071.

Robinson, J.M., Cando-Dumancela, C., Antwis, R.E., Cameron, R., Liddicoat, C., Poudel, R., Weinstein, P. and Breed, M.F. 2021. Exposure to airborne bacteria depends upon vertical stratification and vegetation complexity. *Scientific Reports*, 11(1), p. 9516.

Robinson, J.M., Gellie, N., MacCarthy, D., Mills, J.G., O'Donnell, K. and Redvers, N., 2021b. Traditional ecological knowledge in restoration ecology: a call to listen deeply, to engage with, and respect Indigenous voices. *Restoration Ecology*, 29(4), e13381.

Robinson, J.M., Cameron, R. and Parker, B., 2021c. The effects of anthropogenic sound and artificial light exposure on microbiomes: ecological and public health implications. *Frontiers in Ecology and Evolution*, 9, 662588.

Robinson, J.M., Cameron, R. and Jorgensen, A., 2021d. Germaphobia! does our relationship with and knowledge of biodiversity affect our attitudes toward microbes? *Frontiers in Psychology*, 12, p. 678752.

Robinson, J.M., Aronson, J., Daniels, C.B., Goodwin, N., Liddicoat, C., Orlando, L., Phillips, D., Stanhope, J., Weinstein, P., Cross, A.T. and Breed, M.F., 2022a. Ecosystem restoration is integral to humanity's recovery from COVID-19. *The Lancet Planetary Health*, 6(9), e769-e773.

Robinson, J.M., Mavoa, S., Robinson, K. and Brindley, P., 2022b. Urban centre green metrics in Great Britain: A geospatial and socioecological study. *PLoS One*, 17(11), e0276962.

Robinson, J.M., Harrison, P.A., Mavoa, S. and Breed, M.F., 2022c. Existing and emerging uses of drones in restoration ecology. *Methods in Ecology and Evolution*, 13(9), pp. 1899–1911.

Robinson, J.M., Redvers, N., Camargo, A., Bosch, C.A., Breed, M.F., Brenner, L.A., Carney, M.A., Chauhan, A., Dasari, M., Dietz, L.G. and Friedman, M. 2022. Twenty important research questions in microbial exposure and social equity. *Msystems*, 7(1), e01240-21.

Robinson, J. 2023. *Invisible Friends: How Microbes Shape Our Lives and the World Around Us*. Pelagic Publishing.

Robson, J.P. 2007. Local approaches to biodiversity conservation: lessons from Oaxaca, southern Mexico. *International Journal of Sustainable Development*, 10(3), pp. 267–286.

Rohatyn, S., Yakir, D., Rotenberg, E. and Carmel, Y. 2022. Limited climate change mitigation potential through forestation of the vast dryland regions. *Science*, 377(6613), pp. 1436–1439.

Ronca, S. 2017. Polyethylene. In *Brydson's Plastics Materials* (pp. 247–278). Butterworth-Heinemann.

Rook, G.A. 2012. A Darwinian view of the hygiene or "old friends" hypothesis. *Microbe*, 7(4), pp. 173–180.

Rosenblad, K.C., Baer, K.C. and Ackerly, D.D. 2023. Climate change, tree demography, and thermophilization in western US forests. *Proceedings of the National Academy of Sciences*, 120(18), e2301754120.

Rots, A.P. 2015. Sacred forests, sacred nation: the shinto environmentalist paradigm and the rediscovery of 'chinju no mori'. *Japanese Journal of Religious Studies*, pp. 205–233.

Rull, V. 2019. Human discovery and settlement of the remote Easter Island (SE Pacific). *Quaternary*, 2(2), p. 15.

Rull, V. 2020. The deforestation of Easter Island. *Biological Reviews*, 95(1), pp. 124–141.

Sagan, Carl. *Cosmos, A Personal Voyage*. Chapter II 'One Voice in the Cosmic Fugue'. October 5, 1980.

Salati, E., Lovejoy, T.E. and Vose, P.B. 1983. Precipitation and water recycling in tropical rain forests with special reference to the amazon basin. *Environmentalist*, 3(1), pp. 67–72.

Šantl-Temkiv, T., Gosewinkel, U., Starnawski, P., Lever, M. and Finster, K. 2018. Aeolian dispersal of bacteria in southwest Greenland: their sources, abundance, diversity and physiological states. *FEMS Microbiology Ecology*, 94(4), p.fiy031.

Santos, J.C., de Almeida, W.R. and Fernandes, G.W. 2021. Arthropods: why it is so crucial to know their biodiversity? *Measuring Arthropod Biodiversity: A Handbook of Sampling Methods*, pp. 3–11.

Sapes, G., Demaree, P., Lekberg, Y. and Sala, A. 2021. Plant carbohydrate depletion impairs water relations and spreads via ectomycorrhizal networks. *New Phytologist*, 229(6), pp. 3172–3183.

Scheidt, R.J. 2018. The Babushkas of Chernobyl. *The Gerontologist*, 58(1), pp. 193–195.

Schmidt, K.A. and Ostfeld, R.S. 2001. Biodiversity and the dilution effect in disease ecology. *Ecology*, 82(3), pp. 609–619.

Schnitzler, H.U., Moss, C.F., and Denzinger, A. 2003. From spatial orientation to food acquisition in echolocating bats. *Trends in Ecology and Evolution*, 18, 386–394.

Selden, V. and Putz, F.E. 2022. Root cropping by pocket gophers. *Current Biology*, 32(13), pp.R734-R735.

Shim, R. 2022. Pre-breeding towards the effective utilization of plant genetic resources. *Plants*.

Schultz, C.A., Jedd, T. and Beam, R.D. 2012. The Collaborative Forest Landscape Restoration Program: a history and overview of the first projects. *Journal of Forestry*, 110(7), pp. 381–391.

Shapiro, L. and Stolz, S.A. 2019. Embodied cognition and its significance for education. *Theory and Research in Education*, 17(1), pp. 19–39.

Simard, S.W. 2018. Mycorrhizal networks facilitate tree communication, learning, and memory. *Memory and Learning in Plants*, pp. 191–213.

Simard, S. 2021. *Finding the Mother Tree: Uncovering the wisdom and intelligence of the forest*. Penguin UK.

Singh, C. and Saini, G. Sustainable solution for urban environment: Miyawaki Forest. *International Journal of Technical Innovation in Modern Engineering & Science* (IJTIMES), 5(4), 479.

Skinner, B.F. 2014. Radical Behaviorism. In *An Introduction to Theories of Personality* (pp. 289–315). Psychology Press.

Smith, R. and Botha, J. 2005. The recovery of terrestrial vertebrate diversity in the South African Karoo Basin after the end-Permian extinction. *Comptes Rendus Palevol*, 4(6-7), pp. 623–636.

Smith, L.G., Kirk, G.J., Jones, P.J. and Williams, A.G. 2019. The greenhouse gas impacts of converting food production in England and Wales to organic methods. *Nature Communications*, 10(1), p. 4641.

Soga, M. and Gaston, K.J. 2018. Shifting baseline syndrome: causes, consequences, and implications. *Frontiers in Ecology and the Environment*, 16(4), pp. 222–230.

Son, G.R., Therrien, B. and Whall, A. 2002. Implicit memory and familiarity among elders with dementia. *Journal of Nursing Scholarship*, 34(3), pp. 263–267.

Song, X.M., Wang, J.P., Sun, P.C., Ma, X., Yang, Q.H., Hu, J.J., Sun, S.R., Li, Y.X., Yu, J.G., Feng, S.Y. and Pei, Q.Y. 2020. Preferential gene retention increases the robustness of cold regulation in Brassicaceae and other plants after polyploidization. *Horticulture Research*, 7.

Špulerová, J., Izakovičová, Z., Vlachovičová, M. and Černecký, J. 2022. Natural or Semi-natural Landscape Features as Indicator of Biocultural Value: Observations from Slovakia. *Human Ecology*, 50(3), pp. 531–543.

Squire, D.T., Richardson, D., Risbey, J.S., Black, A.S., Kitsios, V., Matear, R.J., Monselesan, D., Moore, T.S. and Tozer, C.R. 2021. Likelihood of unprecedented drought and fire weather during Australia's 2019 megafires. *Climate and Atmospheric Science*, 4(1), p. 64.

Stamets, P. 2005. *Mycelium Running: How Mushrooms Can Help Save the World.* Ten Speed Press.

Steffen, W., Richardson, K., Rockström, J., Schellnhuber, H.J., Dube, O.P., Dutreuil, S., Lenton, T.M. and Lubchenco, J. 2020. The emergence and evolution of Earth System Science. *Nature Reviews Earth & Environment*, 1(1), pp. 54–63.

Suleri, J.I. 2018. Pro-environmental Printing Behaviour as an Indicator to Reduce Paper Cost at Work. *European Journal of Sustainable Development Research*, 2(3), p. 38.

Sunder, S.S., 1993, October. The ecological, economic and social effects of Eucalyptus. In *Proceedings of the Regional Expert Consultation on Eucalyptus* (Vol. 1). FAO Rome Italy.

Susanti, N.K.Y., Mardiastuti, A. and Wardiatno, Y. 2020, July. Microplastics and the impact of plastic on wildlife: a literature review. In *IOP Conference Series: Earth and Environmental Science* (Vol. 528, No. 1, p. 012013). IOP Publishing.

Susmozas, A., Moreno, A.D., Romero-García, J.M., Manzanares, P. and Ballesteros, M. 2018. Designing an olive tree pruning biorefinery for the production of bioethanol, xylitol and antioxidants: A techno-economic assessment. *Holzforschung*, 73(1), pp. 15–23.

Szepesi, M. 2021. The Effects of Organic vs. GMO Cotton on the Fast Fashion Industry. (November 6, 2021). *The Social Science Research Network* (SSRN).

Tafoya, K.A., Brondizio, E.S., Johnson, C.E., Beck, P., Wallace, M., Quirós, R. and Wasserman, M.D. 2020. Effectiveness of Costa Rica's conservation portfolio to lower deforestation, protect primates, and increase community participation. *Frontiers in Environmental Science*, 8, p. 580724.

Tanacredi, J.T. 2019. Preserving biological diversity: coastal ecosystem restoration not in balance. In *The Redesigned Earth: A Brief Review of Ecology for Engineers, As If the Earth Really Mattered*. pp. 29–52. New York: Springer Publishing.

Tarnovsky, Y.C., Taiber, S., Nissan, Y., Boonman, A., Assaf, Y., Wilkinson, G.S., Avraham, K.B. and Yovel, Y. 2023. Bats experience age-related hearing loss (presbycusis). *Life Science Alliance*, 6(6).

Teo, H.C., Zeng, Y., Sarira, T.V., Fung, T.K., Zheng, Q., Song, X.P., Chong, K.Y. and Koh, L.P. 2021. Global urban reforestation can be an important natural climate solution. *Environmental Research Letters*, 16(3), 034059.

Tian, H. and Liu, X. 2022. Pro-environmental behavior research: Theoretical progress and future directions. *International Journal of Environmental Research and Public Health*, 19(11), p. 6721.

Timmis, K. and Ramos, J.L. 2021. The soil crisis: the need to treat as a global health problem and the pivotal role of microbes in prophylaxis and therapy. *Microbial Biotechnology*, 14(3), pp. 769–797.

Tollefson, J. 2020. Why deforestation and extinctions make pandemics more likely. *Nature*, 584(7820), pp. 175–177.

Torralba, M., Fagerholm, N., Burgess, P.J., Moreno, G. and Plieninger, T. 2016. Do European agroforestry systems enhance biodiversity and ecosystem services? A meta-analysis. *Agriculture, Ecosystems & Environment*, 230, pp. 150–161.

Tree, I. 2018. *Wilding: The Return of Nature to a British Farm*. Pan Macmillan.

Tsien, T.H. 1973. Raw materials for old papermaking in China. *Journal of the American Oriental Society*, pp. 510–519.

Tukey, J.W. 1977. *Exploratory data analysis* (Vol. 2, pp. 131–160). London: Pearson.

Turner, M.D., Carney, T., Lawler, L., Reynolds, J., Kelly, L., Teague, M.S. and Brottem, L. 2021. Environmental rehabilitation and the vulnerability of the poor: The case of the Great Green Wall. *Land Use Policy*, 111, p. 105750.

Uexküll, J.V. 1909. *Umwelt und Innenwelt*. Verlag von Julius Springer. Berlin.

UN FAO Global Forest Resources Assessment. (2020). Online: https://www.fao.org/3/ca9825en/ca9825en.pdf. Accessed on 10.07.23.

van der Ent, R. and Savenije, H. 2010. Length and time scales of atmospheric moisture recycling. *Atmospheric Chemistry and Physics*, 10, pp. 21867–21893.

Vanham, D., Mekonnen, M.M. and Hoekstra, A.Y. 2020. Treenuts and groundnuts in the EAT-Lancet reference diet: concerns regarding sustainable water use. *Global Food Security*, 24, p. 100357.

Van't Padje, A., Werner, G.D. and Kiers, E.T., 2021a. Mycorrhizal fungi control phosphorus value in trade symbiosis with host roots when exposed to abrupt 'crashes' and 'booms' of resource availability. *New Phytologist*, 229(5), pp. 2933–2944.

van't Padje, A., Oyarte Galvez, L., Klein, M., Hink, M.A., Postma, M., Shimizu, T. and Kiers, E.T., 2021b. Temporal tracking of quantum-dot apatite across in vitro mycorrhizal networks shows how host demand can influence fungal nutrient transfer strategies. *The ISME journal*, 15(2), pp. 435–449.

Veldman, J.W., Aleman, J.C., Alvarado, S.T., Anderson, T.M., Archibald, S., Bond, W.J., Boutton, T.W., Buchmann, N., Buisson, E., Canadell, J.G. and Dechoum, M.D.S. 2019. Comment on 'The global tree restoration potential'. *Science*, 366(6463), eaay7976.

Vlot, A.C., Sales, J.H., Lenk, M., Bauer, K., Brambilla, A., Sommer, A., Chen, Y., Wenig, M. and Nayem, S. 2021. Systemic propagation of immunity in plants. *New Phytologist*, 229(3), pp. 1234–1250.

Waller, L.P., Allen, W.J., Barratt, B.I.P., Condron, L.M., França, F.M., Hunt, J.E., Koele, N., Orwin, K.H., Steel, G.S., Tylianakis, J.M. and Wakelin, S.A. 2020. Biotic interactions drive ecosystem responses to exotic plant invaders. *Science*, 368(6494), pp. 967–972.

Wang, W., Feng, B., Xiao, J., Xia, Z., Zhou, X., Li, P., Zhang, W., Wang, Y., Møller, B.L., Zhang, P. and Luo, M.C. 2014. Cassava genome from a wild ancestor to cultivated varieties. *Nature Communications*, 5(1), p. 5110.

Ward, S.F. and Aukema, B.H. 2019. Anomalous outbreaks of an invasive defoliator and native bark beetle facilitated by warm temperatures, changes in precipitation and interspecific interactions. *Ecography*, 42(5), pp. 1068–1078.

Watson, B. 2016. The troubling evolution of corporate greenwashing. *Chain Reaction*, 129(20), pp. 15–00.

Watts, T.W., Duncan, G.J. and Quan, H. 2018. Revisiting the marshmallow test: A conceptual replication investigating links between early delay of gratification and later outcomes. *Psychological Science*, 29(7), pp. 1159–1177.

Wells, G.J., Ryan, C.M., Artur, L., Ribeiro, N., Bowers, S., Hargreaves, P., Fernando, J., Farao, A. and Fisher, J.A. 2022. Tree harvesting is not the same as deforestation. *Nature Climate Change*, 12(4), pp. 307–309.

Werner, G.D., Cornwell, W.K., Cornelissen, J.H. and Kiers, E.T. 2015. Evolutionary signals of symbiotic persistence in the legume–rhizobia mutualism. *Proceedings of the National Academy of Sciences*, 112(33), pp. 10262–10269.

Whillans, A. and West, C. 2022. Alleviating time poverty among the working poor: a pre-registered longitudinal field experiment. *Scientific Reports*, 12(1), p. 719.

Whitlock, J.A. 2011. Inferences of diplodocoid (Sauropoda: Dinosauria) feeding behavior from snout shape and microwear analyses. *PLoS One*, 6(4), e18304.

Williams, M. 2003. *Deforesting the Earth: from prehistory to global crisis*. University of Chicago Press.

Williams, M.I. and Dumroese, R.K. 2013. Preparing for climate change: forestry and assisted migration. *Journal of Forestry*, 111(4), pp. 287–297.

Williams, B. 2018. Reconceptualizing Entrenched Notions of Common Law Property Regimes: Maori Self-Determination and Environmental Protection through Legal Personality for Natural Objects. *Buffalo Environmental. Law Journal*, 26, p. 163.

Wilson, D.S. 1975. A theory of group selection. *Proceedings of the National Academy of Sciences*, 72(1), pp. 143–146.

Wilson, A.D. and Golonka, S. 2013. Embodied cognition is not what you think it is. *Frontiers in Psychology*, 4, p. 58.

Wilson-Henjum, G.E., Job, J.R., McKenna, M.F., Shannon, G. and Wittemyer, G. 2019. Alarm call modification by prairie dogs in the presence of juveniles. *Journal of Ethology*, 37(2), pp. 167–174.

Wiltschko, R., and Wiltschko, W. 2019. Magnetoreception in birds. *Journal of the Royal Society Interface*. 16. doi: 10.1098/rsif.2019.0295.

Wolfe, B.T. 2020. Bark water vapour conductance is associated with drought performance in tropical trees. *Biology Letters*, 16(8), p. 20200263.

Woo, J., Yang, H., Yoon, M., Gadhe, C.G., Pae, A.N., Cho, S. and Lee, C.J. 2019. 3-Carene, a phytoncide from pine tree has a sleep-enhancing effect by targeting the GABAA-benzodiazepine receptors. *Experimental Neurobiology*, 28(5), p. 593.

Woody, J.M. and Phillips, J. 1995. Freud's 'Project for a Scientific Psychology' after 100 years: The unconscious mind in the era of cognitive neuroscience. *Philosophy, Psychiatry, & Psychology*, 2(2), pp. 123–134.

Woudstra, J. and Allen, C., eds. 2022. *The Politics of Street Trees*. Routledge.

WRI (2023). Tracking global tree cover. Online. https://www.wri.org/insights/tracking-global-tree-cover-gain. Accessed on 11.02.24.

Wu, B., Hussain, M., Zhang, W., Stadler, M., Liu, X. and Xiang, M. 2019. Current insights into fungal species diversity and perspective on naming the environmental DNA sequences of fungi. *Mycology*, 10(3), pp. 127–140.

Yanai, I. and Lercher, M. 2020. The two languages of science. *Genome Biology*, 21, pp. 1–9.

Yao, Y., Volchek, K., Lambert, P., Brown, C., Semmler, J., Guindon, A., Dumouchel, T., Mahilrajan, T. and McCall, M. 2014. Environmental impacts of the Fukushima and Chernobyl accidents and their remediation: A review.

In *Proceedings of the 37th AMOP Technical Seminar on Environmental Contamination and Response* (pp. 145–174).

Yokouchi, M., Atsugi, T., Logtestijn, M.V., Tanaka, R.J., Kajimura, M., Suematsu, M., Furuse, M., Amagai, M. and Kubo, A. 2016. Epidermal cell turnover across tight junctions based on Kelvin's tetrakaidecahedron cell shape. *Elife*, 5, e19593.

Yong, E. 2022. An immense world: How animal senses reveal the hidden realms around us. Knopf Canada.

Yu, Y., Zhao, W., Martinez-Murillo, J.F. and Pereira, P. 2020. Loess Plateau: from degradation to restoration. *Science of the Total Environment*, 738, 140206.

Zhou, Y., Smith, S.J., Zhao, K., Imhoff, M., Thomson, A., Bond-Lamberty, B., Asrar, G.R., Zhang, X., He, C. and Elvidge, C.D. 2015. A global map of urban extent from nightlights. *Environmental Research Letters*, 10(5), 054011.

Index